S0-BIF-683

CONSCIENCE, GOVERNMENT AND WAR

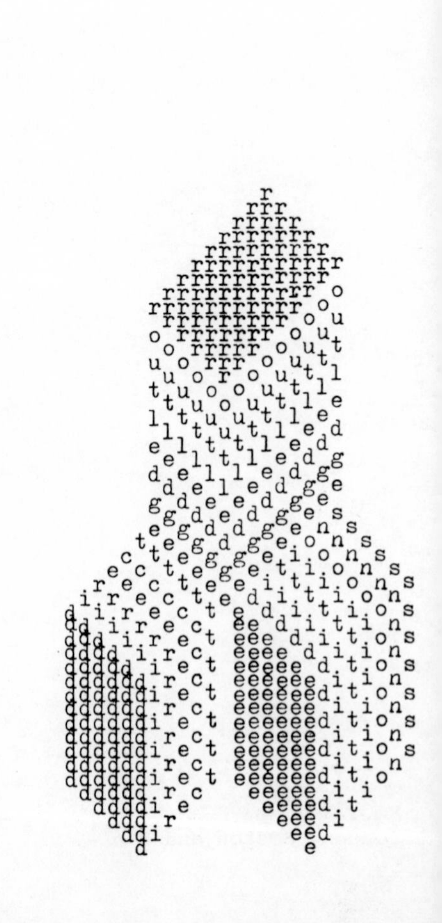

CONSCIENCE, GOVERNMENT AND WAR
Conscientious objection in Great Britain 1939-45

Rachel Barker

ROUTLEDGE DIRECT EDITIONS

ROUTLEDGE & KEGAN PAUL
London, Boston and Henley

First published in 1982
by Routledge & Kegan Paul Ltd
39 Store Street,
London WC1E 7DD,
9 Park Street,
Boston, Mass. 02108, USA,
296 Beaconsfield Parade,
Middle Park, Melbourne, 3206 and
Broadway House,
Newtown Road,
Henley-on-Thames,
Oxon RG9 1EN
Printed in Great Britain by
Thomson Litho Ltd
East Kilbride, Scotland

ISBN 0 7100 9000 5

To my mother and father

CONTENTS

ACKNOWLEDGMENTS

I am indebted to the Social Sciences Research Council for a three-year grant to research and write the dissertation which has formed the basis for this work. I wish to thank Dr Henry Pelling of St John's College, Cambridge, for his consistently constructive and painstaking supervision. I am also grateful to the Master, Fellows and students of Darwin College for making my life as a post-graduate student so happy.

The following people and institutions have kindly granted me access to material in their care: the Keeper of Public Records at the Public Record Office; the Librarian of the University Library, Cambridge; the Librarian at the Trades Union Congress; Edward H. Milligan, Librarian of the Society of Friends who granted me access to both Central Board of Conscientious Objectors and Peace Pledge Union papers, and who, along with his staff, could not have been more welcoming and cooperative; and Miss V.M. West of the Department of Employment, whose prompt responses to all my inquiries was much appreciated. All statistics relating to conscientious objectors are reproduced with the permission of the Controller of Her Majesty's Stationery Office. I would also like to thank those individuals who have talked to me informally of their experiences as conscientious objectors in the Second World War.

The help and understanding of my employers at the British Standards Institution and Sir Fred Catherwood, MEP, during the final stages of preparation have been much appreciated. Lastly, I would like to thank my husband, Ricky, for the unending support he has given me during the years of writing, first, the dissertation and, then, this book.

ABBREVIATIONS

ARP	Air Raid Precautions
AT	Appellate Tribunal
ATR	Appellate Tribunal Report
BM	Board Minutes
CBCO	Central Board of Conscientious Objectors
ECM	Executive Committee Minutes
LAD	Local Authority Decisions
LT	Local Tribunal
LTR	Local Tribunal Report
PC	Press Cuttings
PPU	Peace Pledge Union
PRO	Public Record Office
WO	War Office

INTRODUCTION

Although much has been written about conscientious objection in the First World War, Denis Hayes's book, 'Challenge of Conscience', remains the only full account of conscientious objection in Great Britain in the Second World War.(1) The attention authors of works on the subject have given to the First World War may be accounted for in various ways. Firstly, the Government papers for the Second World War have been only comparatively recently released and, without these, it is difficult to arrive at an accurate and complete view of conscientious objection in that war. Secondly, in 1916 conscription and conscientious objection to military service were new in Britain and it is still fascinating to attempt to answer the questions why a 'conscience clause' was introduced in Britain at all, what form it took, and how it was administered. Lastly, much of the story of conscientious objectors in the First World War is considerably more dramatic than in the Second World War and has therefore been more immediately attractive to historians.

The drama of the First World War was largely the result of abuses of the administration of the 'conscience clause' and consequent injustices and, in many cases, of the violent reaction by the public against conscientious objectors. The purpose of this work is to show whether the British Government used the opportunity to correct these abuses and injustices, whether it was successful in the attempt, and, because in a matter like conscientious objection, the effectiveness of Government action is severely limited if it does not have the general support of society, whether the Government elicited the response from both conscientious objectors and the rest of the populace required to administer the 'conscience clause' fairly.

The legal problems of conscientious objection in both wars have been methodically covered by a Master of Laws thesis by John Hughes, The Legal Implications of Conscientious Objection.(2) Although much of his work is instructive and helpful, the thesis is, rightly, more a work of legal inquiry than a history and has obviously been written for lawyers rather than historians. A slim volume edited by Clifford Simmons entitled 'The Objectors' is a collection of reminiscences of individual conscientious objectors of the Second World War and is interesting and sometimes moving to read.(3) While nothing can detract from the value of individual records of experiences during the war, it cannot be deduced from them that they were in any way typical for the greater part of conscientious objectors of the Second World War. For this reason no accounts of interviews with conscientious objectors have been included in this work although the author has had a number of informal conversations with objectors of the Second World War which have proved most helpful for a clearer understanding of the subject. Pacifism in the Second World War was briefly dealt with by Peter Brock in 'Twentieth Century Pacifism' and the more recently published 'Pacifism in Britain 1914-45' by Martin Ceadel has a very useful chapter on pacifism and conscientious objection in the Second World War.(4)

Hayes's book, then, is the only work which attempts to write a full history of conscientious objection in the Second World War. Hayes was closely involved with the Central Board of Conscientious Objectors, a service organisation for conscientious objectors, and he derived most of his material from the CBCO's records. 'Challenge of Conscience' is a well written and most informative book and the author is indebted to it for all kinds of information and clarification on many points. However, it suffers quite naturally from the fact that it was written so soon after the end of the war and while Hayes was still working with CBCO, which did not disband until conscription was discarded in the 1960s. Hayes makes no attempt to conceal his sympathy with conscientious objectors and a large part of the book describes the particular experiences of conscientious objectors whom Hayes knew and admired. The story is told very much from the viewpoint of the conscientious objector and it has a natural bias towards their interests. Despite this, and despite the absence of Government documents, Hayes manages to write a reliable and accurate account of the history of conscientious objection in the Second World War and this is an admirable achievement.

A gap therefore still exists for a more objective and dispassionate study of this subject and it is hoped that this work goes some way towards filling that gap. Most of the material used is from Government documents kept at the Public Record Office, CBCO and PPU records, multifarious press cuttings held at Friends House Library and from the Parliamentary record, 'Hansard'. There is, in addition, such a wealth of literature on the more philosophical aspects of conscientious objection and on the closely related subjects of the meaning of conscience, justification of civil disobedience in a democracy and, more generally, on the relationship between the individual and the state, that a brief discussion of some of this literature and its relevance to conscientious objection might serve to clear up any outstanding confusion about the exact nature of conscientious objection in the Second World War.

Political philosophers have always of necessity considered the relationship between the state and the individual and especially in time of war. It was to criticise the anarchic state of nature in which continuous war was fought that Hobbes devised his Leviathan. (5) In the nineteenth century, when Government interference in the life of the individual grew with the Industrial Revolution, all kinds of political justification for interference and all kinds of justification for resisting increased interference were propounded.(6)

The advent of the twentieth century heralded an even closer relationship between the state and the individual which culminated in the total commitment to mobilising the whole of society to assist in the war effort in 1914-18. Even those who had roundly supported the concept of 'laissez-faire' now accepted the inevitable growth of state interference. However, there were some thinkers both during and after the First World War who wrote eloquent treatises refuting the right of Government to interfere so greatly with the lives of individuals as to conscript them for military service, but accepting that Governments would always conscript in any future war, devising means whereby war as a means of settling international disputes could be dispensed with altogether. Bertrand Russell, Aldous Huxley and George Bernard Shaw wrote persuasively on this subject.(7) A surge of pacifism between the wars resulted in the production of anti-war literature and more justifications for resisting the state.(8) The growing pretensions of Hitlerian Germany and the apparently unassailable reasons for resisting them put paid to some extent to the pacifist writings of the great political thinkers between the wars.(9) Since then, with the alarming growth of nuclear weapons, more careful, and perhaps less idealistic studies have been made of how the individual should conduct himself in relation to the state in time of war (10) and there is, in addition, a wealth of literature relating to the position of the 'draft dodgers' and conscientious objectors in America during the war in Vietnam.

Much of this literature, however, is concerned with pacifism rather than conscientious objection although this is a distinction few writers have made. An illustration of a confusion which still exists about the two terms can be found in a book written by G.C. Field, published in 1945, entitled 'Pacifism and Conscientious Objection'.(11) Field was a member of one of the conscientious objectors' Tribunals during the Second World War and he set out in his book to refute many of the arguments to which he had listened during the war and to show that there are circumstances in which there are certain evils worse than war which can only be met by fighting. Throughout the book he treated pacifism and conscientious objection as synonymous, an error which throws some doubt on many of his ideas. The distinction he failed to make was that while pacifism is the doctrine or belief that the abolition of war is both desirable and possible,(12) conscientious objection, in time of war, is a term applied to an individual objection which the state accepts as conscientious to some or all of a war effort. Strictly, a conscientious objector is the title or status which the state confers on an individual when it has accepted that the objection to some or all of the war effort is a conscientious one. However, we tend to call all those whose objection is claimed to be on the grounds of conscience, conscientious objectors, whether or not their objection is accepted as conscientious by the state. A 'conscience clause' is incorporated into an Act of Parliament making national service compulsory and any individual is entitled to air his objection before a specially constituted Tribunal which will decide if the objection is conscientious or not. The individual may base his arguments for objection on pacifism, but pacifism is not of necessity the only source of conscientious objection. An individual may make no claims about how others should react to a war or to fighting. He may not, more than any other citizen, consider war evil or wrong, but may consider some part of it unacceptable to his conscience. He may, for instance, find it compatible with his beliefs to help in the war effort as a non-combatant. Such a man is a conscientious objector, not a pacifist. It may be argued that all pacifists should in time of war, automatically be conscientious objectors, but the converse is not true. Field assumed that all applicants appearing before him at his Tribunal were pacifists and when they failed to produce cogent arguments for the doctrine of pacifism or any connections with the well-known pacifist organisations, he was able to record for his book what he thought to be weak and untenable statements in defence of it. The book set out to show the invalidity of the doctrine of pacifism but, for this purpose, used the multifarious and often non-doctrinal arguments for a personal conscientious objection to some or all of the war effort in 1939-45.

This may be a fine distinction but it illustrates the true nature of conscientious objection, that it is not a policy or a doctrine, but merely a status conferred upon an individual by the state while conscription is in force. A conscientious objector is someone who obtained, or at least, sought to obtain, that status. And it is the individual nature of conscientious objection which should be stressed. Individuals objected conscientiously to very different facets of the war and submitted very different reasons for their objections. To ignore the diversity of conscientious objection and to try to impose any uniformity on it is to commit a fundamental error. Perhaps the one certain statement that can be made about conscience is that one person's conscience may dictate very different moral action from that of someone else.

But to attempt to decide what conscience is or constitutes is obviously a matter of some importance if a survey of conscientious objection is to be made, and there is a great deal of disagreement among philosophers on this subject. A distinction has been drawn between the so-called 'traditional' conscience and 'critical' conscience.(13) An appeal to the 'traditional' conscience is, apparently, merely an appeal to a 'gut feeling' or to a voice in the heart. The way to a decision on moral action is not based on reason but is simply obedience to this feeling which often originates in the childhood environment or from an ingrained religious belief or both. It is sensed, rather than known, what is right. 'Critical' conscience, however, is that power apparently given to all of us to reason out the facts of the matter and to come to a conclusion, based on that reasoning, on what our actions should be in any given moral situation. The latter form of conscience is generally regarded as the more reliable of the two.

What relevance does this have for a better understanding of conscientious objection? If these philosophers were correct it would be important for it would mean that those who arrived at their objection through 'critical' conscience better deserved exemption than those who had arrived at it through 'traditional' conscience. However, mistakenly or not, we tend to respect moral decisions arrived at through 'traditional' conscience as much as, if not more than, those arrived at through 'critical' conscience, always supposing the distinction is valid. The person with a religion which forbids the eating of pork has presumably consulted a 'traditional' rather than 'critical' conscience and yet we can all respect him for his refusal to eat pork. Certainly in the Tribunals the Tribunal members much preferred the applicant who had either been a member of a pacifist religious organisation for many years or who came from a family which had raised its children to believe that war was 'wrong' and so their appearance at the Tribunal would seem to have been dictated by 'traditional' conscience rather than 'critical' conscience. Those who had arrived at their position individually without any outside pressure, but through an examination of the moral implications of military service and its demands upon them, found it far more difficult to convince the Tribunals that their objection was conscientious. Those who objected to the particular war of 1939-45, a position which logically would be the best illustration of someone who had carefully weighed up all the facts and reasoned them out, were not exempted at all at the beginning of the war, although later on, after a series of decisions in the Appellate Tribunals, they were. Finally, therefore, both 'traditional' and 'critical' conscience were allowed in the Tribunals.

The more pertinent question is how the Tribunals were able to distinguish between a mere objection to, or dislike of, military service and a conscientious objection, however arrived at. Since it did not make any difference from what form of conscience the objection was arrived at, but was more a question of whether an applicant really believed he was 'right' in objecting, the only method which the Tribunals could use to find out whether an objection was conscientious or not was to test the sincerity with which an applicant put forward his view that he was 'in the right'. How the Tribunals tried to test his sincerity will, it is hoped, be revealed in following chapters.

Lastly, one other concept is often confused with that of conscientious objection and that is civil disobedience. It seems obvious, but the mistake is made, that an individual, by availing himself of a 'conscience clause' in an Act of Parliament, could not, however the term civil disobedience is defined, and this is a controversial subject on its own,(14) be committing an act of civil disobedience. If, however, a person refused to obey the order of the Tribunal and defied the state by, for instance, refusing to undergo a medical examination in preparation for his call-up into the Services, or by refusing to comply with his condition of exemption, then, depending on how civil disobedience is defined, he may well have been committing such an act. He was in any case acting illegally. But acts of civil disobedience, if they are not merely selfish or amoral, are also acts resulting from the dictates of conscience and could be called conscientious objections, whether to military service or to some other law which conflicts with a person's conscience. This is why the terms are so easily confused. As one writer points out, (15) the term conscientious objector, for whose who avail themselves of a 'conscience clause' in an Act of Parliament making military service compulsory, may be misleading since it implies that compulsory military service is the only civil law or rule to which it is possible to object. This is, of course, not true but it is no excuse for so confusing the issue that the term conscientious objection is not properly understood in its common usage.

Despite these difficulties the peripheral literature around the philosophical aspects of conscientious objection is of great interest. The teaching of the main churches on man's responsibility towards the state in time of war is particularly fascinating for it often conflicts with the reasons applicants, who based their objection on religious beliefs, gave for their objection.(16) But by whatever inward struggles the aspiring conscientious objector suffered in coming to his moral decision, it now falls to study in detail his experience in the Tribunals, and his experience in whatever part of a society grimly committed to warfare he afterwards found himself, whether his application was accepted or rejected.

Chapter 1

PROSPECTS OF WAR: inter-war pacifism and the introduction of conscription

The urgent desire to settle international disputes without war, which was almost universal in Britain in the 1920s and 1930s, found expression, at least until the mid-1930s, in adherence to the League of Nations and the support of the League of Nations Union.(1) But this popular support (the Union had all party favour) was not all pacifist in a strict sense. The 1934 Peace Ballot, organised by the Union, while prompting an 85 per cent 'Yes' vote for economic measures against an aggressor nation, also secured a 58 per cent 'Yes' vote for the application of military measures. Only 20 per cent of those balloted gave a definite 'no' to military measures.(2) On the other hand, when Canon Dick Sheppard published a letter in the national press on 16 October 1934, appealing to men with pacifist ideals to let themselves be known to him, he and his supporters were astounded (after a disturbing few days when there were no replies at all because the Post Office was holding them back for a large single delivery) to receive 2,500 replies within two days, far more than expected. The number eventually increased to 50,000.(3)

A significant minority of the population, therefore, believed itself to be pacifist while the great majority of the remainder was pacifist only in its revulsion against war and its determination to avoid it if at all possible. In this endeavour it was well represented by the National Government of the 1930s, while the parties of opposition, starting out the decade with at least partly pacifist policies, gradually came to assume a similar position. In 1933 the Labour party conference had supported a motion by Sir Charles Trevelyan pledging the Labour Movement 'to take no part in war and resist it with [its] whole force', and to consider what steps, including a general strike, might be taken to prevent it. The leader of the party, George Lansbury, was a convinced pacifist but there was also support from those socialists who believed war to be a clash of rival imperialist claims arising out of the contradictions of capitalism. Professor Harold Laski and Sir Stafford Cripps led this group.

In 1934, however, the party's line had altered: peaceful ways of ending disputes had to be sought but the party would support the Government if it had to restrain an aggressor by collective security. In 1935 Lansbury was forced to resign after the party conference had passed overwhelmingly (by 95 per cent) a motion supporting the League of Nations's authority in the Abyssinian crisis.(4) Ernest Bevin is credited with much of the success in persuading the Labour Movement to move away from strict pacifism as a policy. Hector McNeil MP later wrote of Bevin's role: 'His speech [at the party conference] did more than any other to substitute the forward looking leadership of Attlee for the pacifism of Lansbury.'(5)

The quasi-pacifism of the majority of the population and of all the political parties was something rather more, however, than the pacifism all liberal-minded people today claim, usually before statements on the need for preparedness for war. In the 1930s, as yet Great Britain had not fought a war against a commonly accepted unarguable moral evil. The fighting of the Great War had not only been the more horrific of wars, it had also been a war in the old tradition. An aggressive power had disturbed the finely balanced power structure in Europe. Britain had fought, at the cost of countless dead and injured, to restore that balance and that cost seemed to many to outweigh the benefits of 'victory'. And while most were prepared, as a last resort, to enter that kind of war again (as they believed they were doing in 1939), it was an easier kind of war to reject than the kind of war when the enemy is inherently evil as (it emerged) he was in the Second World War. Quasi-pacifism today remembers only a war which the great majority believed was worth fighting, a truly 'just' war. The tendency today is to see all future wars in that mould. Consequently pacifism can be dismissed as understandable but unrealistic wishful thinking. Moreover, it is still very much in the minds of our leaders that the quasi-pacifism of the 1930s led to military unpreparedness for Hitler's war, a line of argument often used to defend increases in armaments today.

These differences account for the widespread acceptance of a form of pacifism in the 1920s and 1930s not only among the public but also among politicians and other public figures, many of whom later reversed their positions when the nature of the enemy in the new war became clear to them. A.A. Milne wrote in 1940 that there was one reason only why he had changed his mind since writing the pacifist 'Peace With Honour' in 1934 and that was, simply, Hitler. He had rejected war as he had known it, that is as competitions between countries to secure material advantage, but Hitler was threatening 'all Christian and civilised values', and had to be resisted. He wrote, 'I am still a pacifist but I hope a practical pacifist. I still want to abolish war.'(6) Few would have disagreed with that. Cyril Joad and Bertrand Russell, who had been a Sponsor of Dick Sheppard's Peace Pledge Union, were among others who felt similarly. (7) The loss of membership of the PPU in the darkest days of 1940 and the downturn in the numbers of conscripts registering as conscientious objectors at the same time are evidence of a significant break with inter-war pacifism. Some well-known conscientious objectors of the Great War had already rejected their previous position: Herbert Morrison, Minister of Home Security in Churchill's Government, is the leading example.(8) F.W. Pethick Lawrence was more consistent in his approach, though. He had admitted a political objection in 1918 when he was called before the Local Tribunal in Dorking stating that he was 'not prepared to say that I was against war in any circumstances which seemed to me to justify a resort to arms.' Exceptionally, his objection was accepted and he was exempted on condition that he undertook land work. His position in the Second World War was therefore logical; the circumstances he felt justified taking to arms had arrived.(9)

However, a mass pacifist movement which outlasted the war years did take shape in the 1930s. Canon Dick Sheppard, who had been Vicar of St Martin's in the Fields where he had built up an enthusiastic following and was now a Canon of St Paul's Cathedral, encouraged by the support for his letter in the national press, held a rally in the Albert Hall, on 14 July 1935, which over 7,000 attended. Other speakers included Edmund Blunden, Siegfried Sassoon, the novelist Maude Royden and Brigadier-General Frank Crozier, the war veteran whose pacifist origins were ambiguous but who had 'fallen under Sheppard's spell' and worked tirelessly for Sheppard and for pacifism.(10) The PPU was launched in May 1936 at a meeting of Sheppard, Arthur Ponsonby, the novelist Margaret Storm Jameson, the Revd Donald Soper, J.H. Judson, Canon Charles Raven and Crozier.(11) Sheppard's influence on the Union was immeasurable and his death six months later was a blow to the cohesion of a body which, because it attracted people with such strongly held views, not always in complete harmony, was difficult to lead. The mass popularity of the PPU, which Sheppard had inspired in 1934 and institutionalised in 1936, lessened after his death, although there was still a steady flow of new pledges. Munich, and then the advent of conscription, however, brought a surge of recruits.(12)

That conscription, or talk of conscription, should increase membership of the PPU is not surprising. It was first introduced in 1916 amid stormy parliamentary scenes and had been dispensed with in 1919 after the Great War. Throughout the 1930s, as the international situation deteriorated, governments, starting with Baldwin's in 1936,(13) had made repeated assurances in Parliament that the much disliked conscription would not be re-introduced in peace time. The country might rearm (Baldwin succumbed to pressure for a Ministry of Supply in March 1936 when he appointed Sir Thomas Inskip as Minister for the Coordination of Defence, although Inskip was a 'routine lawyer' and a 'nonentity' and had to be 'prodded' in Parliament by Churchill (14)), but while there was no conscription there could be hope that there would be no war. Already in 1936 there was pressure from Britain's allies to conscript as well as to rearm. In that year Leon Blum visited Britain and urged his fellow socialists not to oppose the Government's rearmament programme and actively to press for conscription if it was the only way the necessary forces could be raised. Ernest Bevin hardly needed persuading and continued to press the Labour Movement to repudiate the pacifist tendency. By 1937 both the trades unions and the Labour party had withdrawn overt opposition to rearmament and both TUC and Labour party conferences in the autumn of 1937 adopted, with overwhelming support, a report on foreign policy by the National Council of Labour which made clear that unless conditions improved 'a future Labour government would not reverse the policy of rearmament.'(15) The need for conscription was not yet accepted but it was only a question of time. Its position on the issue was becoming, apart from the rhetoric, not dissimilar to that of the Conservative party - conscription would have to come, but had to be put off until the last possible moment, for it signified failure to avert war.

In fact a conscription bill (with provision for conscientious objection) had been in readiness since the last war, as the unfortunate Sir Thomas Inskip revealed almost by chance in Parliament in a debate on the coordination of wartime service enrolment, causing a public outcry which the national press, especially the 'Daily Mail', took up.(16) Chamberlain made his displeasure at the disclosure evident when, two days later, he replied to Opposition questioning, 'I think myself that if my right hon Friend had had longer notice with which to prepare his words he probably would have expressed himself somewhat differently.' What Inskip 'meant to say' (17) was that volunteers of the armed and civil defence services could not be certain, if war broke out, that they would not be compulsorily transferred to other departments. The row confirmed what was already quite clear to Chamberlain: conscription was still very much a live political issue.

Meanwhile there were continued pressures on the Government at least to raise more men by voluntary methods (L.S. Amery was a persistent parliamentary critic (18)) and in the autumn of 1938 Sir John Anderson was made Lord Privy Seal in charge of civil defence. The National Service Campaign held its inaugural rally in January 1939 in London, where platform speakers included Anderson, Ernest Brown, the Minister of Labour and Herbert Morrison. A pamphlet went to every household in the country, persuading, rather than threatening, householders to join up. National Service Committees promoted recruitment locally, coordinated by a Central Committee presided over by the Earl of Onslow. Two million applications were made during this time out of which there were one-and-a-half million enrolments. Three-quarters of this figure volunteered for civil defence, an indication of the defensive mood of the country.(19)

The Labour Movement and the trades unions were suspicious of the campaign. Six of the larger trade unions refused to co-operate with it and some local Labour parties were 'restive'. But a large majority was more concerned with safeguards: that there should be no conscription of labour without 'conscription of wealth' and that there should be no industrial conscription. The PPU issued Peace Service Handbooks in response to the campaign, 'a guide suggesting some of the ways by which the people of Britain can help their country and the world to live at peace.' Some 200,000 copies were sold or distributed.(20)

Despite John Anderson's efforts, the numbers raised for the Armed Forces, as with the Derby Scheme of the First World War, were insufficient. Internal pressures for the introduction of compulsory military service mounted from the Armed Forces and from Conservative backbenchers, some of whom belonged to the Citizen Service League, formed also in January 1939, specifically to promote conscription. The Marquess of Salisbury was a vice-chairman and L.S. Amery, the bishop of London, and Field Marshal Lord Milne were on the Council.(21) Lord Vansittart, who was then diplomatic adviser to the Foreign Secretary, became so desperate in the spring of 1939, that he privately begged prominent French leaders to inform the Government frankly that they could no longer guarantee the Entente unless Britain immediately introduced conscription.(22)

And, indeed, the French did put pressure on the Government to act when the president of France visited Britain in March 1939.(23) Chamberlain was acutely aware of the pressures but also of the pledges he had made that conscription would not be introduced in peace time. In 1938, in Cabinet, the Prime Minister had (24)

> agreed that public opinion was moving rapidly towards putting defences on a more sure footing... He himself had been careful not to use language inconsistent with some form of compulsion [but] the moment had not yet arrived.

But in March 1939,

> He had a great deal of sympathy with those who said that there was nothing which would so impress foreign opinion as the adoption by the country of compulsory military service.

He had by this time arrived at the conclusion that the truth of his pledges would not be prejudiced, for the spring of 1939 'could not be described as peace time'. It was not his pledges that now held him back, but his concern about the possible reaction of the Labour party and the TUC. At the moment they were being co-operative in the rearmament programme and in the voluntary recruitment campaign, but the introduction of compulsory military service might upset the delicate relationship between Government and the Opposition. From 'soundings' Chamberlain had made, he felt it would be 'a very dangerous step'. The Cabinet concluded on this occasion that an increase in voluntary recruitment was needed.(25)

Sensitivity to the reactions of Labour was again demonstrated in the memorandum before the decisive Cabinet meeting on 24 April, which indicated the lines upon which a measure of compulsory military training might be arranged. It was 'designed to avoid as far as possible drawing the opposition of the Labour Movement.' At that meeting it was agreed that international pressures to introduce conscription, now from the USA as well as France, were becoming overwhelming. Chamberlain was still preparing his case for breaking his pledges: his undertakings 'did not disturb him' as these were not times of peace. The introduction of a bill in the form of 'exceptional powers' would pacify trades unions who had always made their feeling known that they would reject compulsion as part of ordinary peace-time practice. The powers would be presented as supplementing, rather than superseding, voluntary methods. Chamberlain told the Cabinet that he wanted to give prior notice to trades-union leaders before the measure was announced in Parliament. The decision had been made.(26)

On the morning of 26 April representatives of the TUC were summoned to 10 Downing Street and told of Her Majesty's Government's intention to introduce a limited form of conscription for men aged 20 and 21. The Prime Minister announced the news to the House of Commons that afternoon. Chamberlain had known for some time that conscription was inevitable, but the final decision was forced by internal and external pressures and by Chamberlain's last hope - that what was for Britain a grand gesture like this would deter the enemy - 'a final earnest of Britain's purpose'.(27)

Chapter 2

THE TRIBUNALS: their procedures and decisions

Chamberlain had judged correctly that public opinion was ready for the introduction of conscription; its announcement was received with only half-hearted opposition. Attlee accused Chamberlain of breaking his pledge,(1) as Chamberlain had expected he would, and John McGovern (Labour) called for a general election or a referendum in view of the broken pledges,(2) but in general the debate in Parliament was more concerned with the bill's detailed provisions than with arguments for and against conscription.(3) Lack of mass opposition, both in Parliament and in the country, was partly because it had been so long expected, despite the pledges, partly because it was only a small measure of conscription affecting 31,000 young men per annum as Chamberlain estimated,(4) and partly because the gravity of the European situation in any case seemed to warrant it.

As the Bill introducing conscription, made statutory in May 1939, invited little public comment, neither did the inclusion of a 'conscience clause'. The precedent of the 1916 Military Training Act also made this unremarkable. But while the statutory provisions for conscientious objectors were based on those in the First World War, significant changes were made, and these were largely in the membership and procedure of the Tribunals and their role in a society at war. The main difference lay in the fact that in the Second World War the Tribunals were set up specifically to deal with applications from objectors to be exempted from some part of the war effort, and were called Conscientious Objectors' Tribunals. They were administered by the Ministry of Labour and National Service,(5) and their membership was strictly controlled and vetted by that Ministry. In the First World War, however, the Tribunals developed from those which administered the Derby voluntary recruiting scheme; they were under the aegis of the War Office, until 1917 when all recruiting matters were put under the control of the Ministry of Labour and National Service,(6) and composed of local dignitaries, often from the local Council, who still saw their role as recruiting officers. Moreover, their job was not only confined to dealing with conscientious objectors; they dealt with hardship cases as well, that is those who applied to be exempted from military service for medical or personal reasons. A military representative was allowed to be present and in many cases acted in the manner of a prosecutor in a court case. There was very little liaison between the Tribunals and minimal centralised control. While many First World War Tribunals managed to carry out their duties with reasonable efficiency and fairness under the circumstances, there were obvious anomalies from which lessons could be learnt.

CONSTITUTION, FUNCTIONS AND PROCEDURE OF LOCAL AND APPELLATE TRIBUNALS

The Act which introduced full-scale conscription, the National Service (Armed Forces) Act 1939, was introduced on 1 September after the German invasion of Poland and when war was quite inevitable. The provisions for conscientious objectors were identical in both Acts of Parliament. The section relating to objectors,(7) stated that if a man required to register under the Act, conscientiously objected to being placed on the military service register, or to performing military service or to performing combatant duties, he should apply for his name to be placed on the register of conscientious objectors. The man could then make an application to a Local Tribunal stating to which matter he objected. If satisfied that the application was genuine, the Tribunal could register the applicant on the register of conscientious objectors without conditions, that is complete exemption, or register him on condition that he undertook civilian work under civilian control, often land work, especially at first, or order that his name should be removed from the register but that he should be called up for non-combatant duties only in the Armed Forces. If not satisfied that the application was genuine, the Tribunal could remove the name without condition from the register. The person concerned would then be called up to the Armed Services. If the applicant was dissatisfied with the decision at the Local Tribunal, he could take his application to an Appellate Tribunal where the decision was final; none of these decisions could be called into question by a court of law.

Local Tribunals were to consist of a chairman and four other members, one of whom should only be appointed after consultation with a trade-union organisation,(8) and the Minister was to have regard 'to the necessity of selecting impartial persons'. The chairman was to be a County Court Judge,(9) or in Scotland a Sheriff. An Appellate Tribunal was to consist of a chairman nominated by the Lord Chancellor, or in Scotland by the Lord President of the Court of Session, and two other members.(10) Reasonable notice of the time and place of the hearing was to be given to the applicant although the hearing could proceed if the applicant were not present, and the hearings would be heard in public unless for any reason the Chairman should think otherwise. The applicant could be represented by a friend, relative, trade-union official or by counsel or solicitor. The decisions by the Local and Appellate Tribunals could be made in private and the Chairman of the Tribunal had the power to administer the oath. The decision of the majority would be the decision of the Tribunal although, if the decision was not unanimous, this would have to be recorded in the report of the proceedings of the Tribunal. Aside from these regulations the procedure of a Tribunal was to be as the Chairman determined.(11)

England and Wales were divided into 11 districts and Scotland into 4,(12) and a Local Tribunal normally sat at a town where there was a divisional office of the Ministry of Labour and National Service. The Tribunals could consider cases, referred to them by the Minister, of persons who had refused or failed to register provisionally on the register of conscientious objectors, but who appeared to the Minister to be conscientious objectors. For instance, in April 1940 the Minister of Labour presented two applicants at the Manchester Local Tribunal who had failed to register on the register of conscientious objectors because 'there were reasonable grounds for considering them conscientious objectors.'(13) Tribunals could also consider fresh applications by those who had failed to observe a condition imposed by a previous order of the Tribunal.

Any officer of the Ministry, approved by the Minister, could be heard at the Tribunal, and the Minister could be represented by counsel or solicitor. A clerk to the Tribunal, who was also an officer of the Ministry, was to be appointed for clerical duties, but could take no part in the proceedings of the Tribunal. When the Northern Divisional Office informed the Ministry of Labour in London that it had combined the duties of the Minister's representative with those of the clerk in the Northern Tribunal, the Ministry warned that this was highly inadvisable, for the two jobs were entirely different. The clerk could offer factual information but could never advise. If economy of manpower was the reason for this action, the Ministry felt obliged to point out that the Minister's representative need not be a First Class Officer (who would be more highly paid) and also need not attend every sitting.(14) The decision of the Tribunal was to be notified to the applicant by means of a signed order which was to be handed or posted to the applicant. Both the applicant and the Minister could appeal to the Appellate Tribunal, of which there was one for England and Wales, and one for Scotland, and whose decision was final.(15)

Some newly appointed Tribunal Chairmen were worried at first about the provisions. Judge Hargreaves, who was appointed Chairman of the Fulham Local Tribunal, wrote to the Ministry expressing his concern about several points. He wondered, for instance, how long it would take the Tribunals to deal with the 3,300 applications outstanding by mid-1939; he thought 20 minutes would be a reasonable time to deal with each case, but asked that someone should check the records of the First World War to find out how long their cases had taken. His next point was to express concern over the venue of the Tribunals. He suggested that outside London, they should sit at County Courts, as these courts did not sit every day; they could also perhaps use Registrars' Courts in the afternoons or local Town Halls. He said that his own County Court was free from the 19 June and that it was 'nice and cool'; he could work there for a whole week. He considered that there would be no need for the Chairman to take notes because the case could not recur and that an appeal would be as good as a rehearing. He thought that a short note might be necessary if the applicant was refused, giving the reasons for the refusal. He also asked whether there would be any rules of procedure and supposed 'some would be necessary'. Again he stressed the importance of studying the experience of the First War. He felt that applicants who were refused should not have to join the six-months' training course for conscripts half-way through, as they would then be 'marked men', but they would have to do all the training at some time. He asked whether it would be suitable to inquire of the applicant which alternative occupation he would prefer to undertake.

The Judge then telephoned the Ministry of Labour asking further questions. He wanted to know when his Tribunal was likely to start work, whether he should wear robes, whether a representative of the Ministry would be a member of the Tribunal, and whether Tribunals would have to sit during August. The Ministry suggested that the Judge should do as he thought best, pointing out, however, that ordinary court cases would have to continue, but that while this would cause a delay in the hearing of cases, conscientious objectors would not suffer any hardship since they could not be called up until after the hearing was over. They told the Judge that the hearing of cases would not start until the middle of July and that August holidays would not be affected. A representative of the Ministry would be present, but not in a legal capacity. A clerk would be provided. The Lord Chancellor's department would decide where the sittings should be held and whether robes should be worn. A permanent place of sitting would be found. No general rules were thought to be necessary.(16) By the time the National Service Act had received the Royal Assent, however, these questions had been settled, and explanatory notes for the Chairmen and members cleared up any outstanding confusions.

THE FIRST MONTHS

The pressure of work on the Tribunals at the outset of the war was enormous. A Glasgow newspaper commented in mid-1941 that the Glasgow tribunal was at last up to date with its work.

> At the outset, applicants were so numerous that men had to wait 6 or 7 months before they were called before a tribunal. Many men, seeing in this an opportunity for at least a temporary reprieve, appealed without genuine conviction and without hope in succeeding. In the early days, the percentage of 'conchies' among men registering was 1.5. In the latest age group, 39's, it has fallen to 0.5.

The paper said that a Ministry of Labour official had commented that the number registering as objectors had been falling as each historic event, such as the Fall of France, had occurred, and that he thought the entry of Russia into the war would almost certainly affect the numbers.(17) In this newspaper article are two implied explanations for the rapid fall in the number of persons registering as objectors as the registrations progressed. Firstly, the author suggests that men, knowing they could not be called up until after the Tribunal hearing, and knowing also that there were severe delays in the hearings of cases, took advantage of this information to put off the 'evil day' when they would have to enter one of the Armed Forces. The idea seems plausible enough on first sight, but the figures do not bear it out. The percentage of persons whose applications were regarded by the Tribunals as not genuine was much lower at the beginning of the war than at any other time, even though the number registering was higher. Numbers registering provisionally fell rapidly as the war progressed,(18) but the percentage of those applications rejected by the Tribunals rose quite sharply. Therefore all those 'extra' applicants at the beginning of the war were not regarded by the Tribunals as impostors. Perhaps the second and better explanation that the article offers is that the age groups required to register at the beginning of the war were the younger age groups, the twenties. Young men were more likely to be genuinely idealistic especially as, at the beginning of the war, it was not totally clear for what issues Britain was fighting. The events of the first part of the war, as the Ministry of Labour officials in the newspaper pointed out, would probably have persuaded many older men that they must forsake their pacifist principles to fight for even more important principles which the German expansion appeared to threaten. In any case the Tribunals were certainly overloaded with work at the beginning of the war, but when they managed to get up to date by 1941, provisional registrations had fallen to a manageable level.

Although the Act and the guidance notes issued by the Ministry of Labour clarified the constitution, functions and procedure of the Tribunals, the manner in which they were to reach their decisions was largely left to the Chairmen and members to decide, and some Chairmen took it upon themselves to define more clearly what they considered to be the Tribunals' role. Judge Wethered of the South-Western Tribunal complained in 1942 that nowhere in either the 1939 or 1941 National Service Acts was the term 'conscientiously objects' defined. He went on to say, 'we have defined for our own guidance, a conscientious objection to military registration or service, based on religious or ethical convictions honestly held.' (19) He thought this definition avoided having to discover whether the objections themselves were reasonable which, in his opinion, was difficult to ascertain. To find out whether the objection was honestly held, 'we have found that the most important factor to consider was the religious or ethical background behind the objection.'(20) If an applicant came from a background, religious or moral, which had strong pacifist connections, and with which he had associated himself for some time, he would be considered to hold his convictions honestly. A long personal history of pacifism was, more or less, a prerequisite for a successful applicant before Wethered's tribunal.

Another Chairman had by 1940 reached certain conclusions over the role of Tribunals. When making a speech thanking well-wishers on the resumption of his place as Chairman of the North-Western Tribunal (after an absence caused by a physical attack on him by an applicant whom the Tribunal had removed from the register of conscientious objectors), Judge Burgis was at pains to point out that, despite accusations otherwise, the Tribunals did not exist to recruit soldiers.(21)

> We are a judicial body. We receive no directions. It is immaterial to this Tribunal how many we send [into the Armed Services]. Conscience and conscience alone is what we have to consider .... All we have to consider is whether these views are sincerely and deeply felt, and we have not to determine whether they are reasonable or patriotic at this juncture.

Judge Burgis had felt it necessary at the outset to define the intentions and proposed role of his Tribunal. At the opening session in November 1939 he announced that he would like to say a few words so that both applicants and members could begin on what he called 'a friendly and understanding basis of approach'.(22)

> My colleagues and I enter upon our duties with feelings of sympathy. We realise the sufferings that war entails .... We realise also the mental anguish that war brings in its train.... We have to ascertain what is in the minds of the applicants, to appraise the genuineness and sincerity of their views, to plumb the depths of their convictions. We have also to see that conscience is not made a cloak.
>
> Now, we can only ascertain whether there is a genuine conscience and a deep conviction by getting to understand the background of the lives of each of those who come before us. In order to understand the background, we must probe by question and answer. And we hope that those who come before us will not resent our questioning or think that we are persecuting conscience .... I hope that the applicants will keep these points before them. If they do they will save themselves much irritation and annoyance, and they will help to promote the sympathy and goodwill with which the Tribunal approach their task, and, above all, they will help us to judge righteously.

Despite these admirable intentions the North-Western Tribunal aroused some criticism in later years, and this will be discussed in the next chapter.

Some Tribunal Chairmen and members looked to the Appellate Tribunals for guidance and complained that they found very little. Judge Wethered in his review of the work of the South-Western felt that in the early days the Tribunal allowed too many unconditional exemptions but blamed this partly on the lack of guidance from the Appellate Tribunals. He felt that their decisions were inconsistent and confusing.(23) And in a letter to the Ministry of Labour, Judge Davis, Chairman of the South-Eastern Tribunal, asked that the decisions of the Appellate Tribunals, which related to the construction of the Act should be made known to the Local Tribunals. The press was the only existing source from which he could derive information.(24)

Another problem which emerged in the first months of the work of the Tribunals was that there was a proportion of Welsh conscientious objectors whose first language was Welsh. It seemed only fair that they should be able to plead their cases in their own language. Sir Artemus Jones, chairman of the North Wales Local Tribunal, had much to do with urging the Minister of Labour to allow Welsh objectors the opportunity to present their cases in the Welsh language. (25)

THE DECISIONS OF LOCAL TRIBUNALS

Unconditional exemption

Of the decisions that the Tribunals could make, the one that proved to be the most controversial, both in and out of the Tribunals, was that of unconditional exemption. The experience of the First World War had shown that a number of objectors would claim that to take any part in the war effort at all would be violating the dictates of their conscience. Tribunal members found this position the most difficult to understand. They could respect and sympathise with an objection to the act of killing or even an objection to being involved in the non-combatant part of the military effort, but what they considered to be a total rejection of all responsibility of the individual to the state, especially in an hour of desperate need, stretched their tolerance to the absolute limit.

There was no doubt that Tribunals had the power to grant exemption from military service without conditions; it was written in both the Military Training Acts of 1939. The Prime Minister, Neville Chamberlain, had referred to this type of objector in his speech at the Second Reading of the Military Training Bill in 1939: (26)

> There is the most extreme case, where a man feels it his duty to do nothing to even aid or comfort those engaged in military operations .... Probably that is the smallest of all classes of Conscientious Objectors. But it often happens that those who hold the most extreme opinions hold them with the greatest tenacity. We learned something about this in the Great War, and I think we found that it was both a useless and an exasperating waste of time and effort to attempt to force such people to act in a manner which was contrary to their principles.

The Government and Parliament therefore adopted a conciliatory tone to which it was hoped, especially by organisations representing objectors, the Tribunals would adhere. But undoubtedly the treatment the Tribunals gave to those who claimed absolute exemption produced by far the most comment and criticism from objectors, the press and the public.

Part of the difficulty was that each Tribunal had different ideas about how the cases of persons asking for unconditional exemption should be examined and treated. The Central Board of Conscientious Objectors (hereafter CBCO) was worried about reported statements by Judge Drucquer of the Southern Local Tribunal that the Tribunal was going to grant no unconditional exemptions at all. In 1943 the Public Relations Officer of the Board wrote to the Judge saying that he had been concerned that the Judge had been reported as saying to an applicant, 'We are not empowered to give you complete exemption.' It was very serious, said the letter, to ignore the law. The CBCO had already written to the Ministry of Labour complaining about the Chairman's reported statement but had received no satisfactory reply (27) and so it was agreed by the Executive Committee in January 1943 that more evidence must be collected proving the Tribunal's bias against unconditional conscientious objection, and this could be done through Local and Appellate Tribunal reporters working for the CBCO or other related organisations.(28) Certainly Judge Drucquer seemed confused as to what he was actually empowered to do and what he was not. If he really had not intended to give any unconditional exemptions, he would have done better not to have announced it publicly. Other Tribunals aroused little controversy even though they consistently refused applications for unconditional exemptions, because they never explicitly suggested this was policy; merely the lack of genuine applications.

The Fulham Local Tribunal in London was one of the least flexible in its attitudes to unconditional exemptions. A reporter for the Peace Pledge Union (hereafter PPU), who listened to all the cases, reported that he could not understand the mentality of the Tribunal. The members were fixed in their beliefs, rigid in their outlook, and unshakable in their determination that everyone should at least do something.(29) Another reporter at the same Tribunal, amazed at the rejection of an applicant who seemed to him to have a very good case, commented, 'Perhaps the tribunal had pre-determined that there should be no total exemptions in today's hearings.'(30) Even when the case of an obviously sincere applicant was supported by a Methodist minister, his request for unconditional exemption was refused; the 'Barnet Press' reporter, possibly a less biased source than the CBCO or PPU reporters, was sufficiently moved to report this case.(31) By May 1940, the CBCO reporter noticed that there was a definite hardening of attitude among the members of the Tribunal, and there was no intention of granting any full exemptions. (32)

The South-Eastern Tribunal, another well-documented and reported Tribunal, was thought at first by different reporters to be fairer and more just than that at Fulham. In November 1939, a CBCO reporter noticed that, while applicants who were asking for unconditional exemptions were questioned closely, the questions were always fair and to the point. If an applicant said that he felt he could not help the war effort in any way, he was always asked how far his present job was already doing this. As long as the applicant was not working in an obviously war-helpful industry like that of munitions, and as long as he had obviously considered this problem, the application would normally be granted.(33) Comparing this Tribunal with that of Fulham, a reporter commented that the South-Eastern Tribunal was so 'tolerant, sympathetic and impartial', that one would hardly had thought the two bodies had the same job on hand. (34) Yet when Judge Hurst was in the chair, the whole atmosphere of the Tribunal appeared to change. In June 1940, a reporter noted that the applicant's statement was often ignored, and that the Tribunal was reluctant to award unconditional registrations to objectors. 'One applicant told me he saw him (Hurst) strike out this clause on the form before he had asked a question.'(35)

Undoubtedly, the decision to grant unconditional exemptions presented special difficulties to the Tribunals. Judge Wethered of the South-Western Tribunal looked to the Appellate Tribunals to give guidance on this point but all that he could find was an appeal decision which said that an applicant who objected to performing military service of any kind should undertake civilian work under civilian control, 'unless the objector proved that he had a well-founded objection to undertaking civil work under civilian control as an alternative to Military Service.' Wethered rightly pointed out that there was no indication in this decision as to exactly what grounds were 'well-founded'. The matter again was left to the individual discretion of the Chairmen and members.(36)

According to the CBCO, there were several factors which aided the case of applicants for unconditional exemption. The reporter at Fulham noted that those applicants who were physically incapacitated were often awarded unconditional exemption even when they had not applied for it.(37) Another reporter at the same Tribunal noted that one applicant had three letters from members of the Armed Forces supporting him, and was registered unconditionally. In a Tribunal which rarely gave this decision, 'Does not this decision suggest that it is a great asset to have letters from members of the fighting forces?'(38)

Of those applying for unconditional exemption, a considerable proportion were known as political objectors, mostly socialists or communists, who did not object to war as such, but who objected to a capitalist war, although the entry of Russia into the war in 1941 confused this issue. Of all objectors it was perhaps this type that was most unpopular in and out of the Tribunals. In 1940, a deputation headed by Dr Alfred Salter MP and by Arthur Creech Jones MP complained to the Secretary of State at the Ministry of Labour and National Service that political objectors' applications were being rejected out of hand. The Ministry's view appeared to be that a political objection could be a genuine conscientious objection, but only if the objection was to all war, and not just to one. The Ministry did not agree with Local Tribunals which granted unconditional exemptions to political objectors on the basis that the objector held his view so intensely that he could be said conscientiously to object. Its view was that it was not the degree of intensity with which the conviction was held, but the matter to which the applicant objected that was the determining factor. However, as always, the Tribunals were free to interpret this as they thought fit. But the Ministry stressed that they did not think people ought to be allowed to 'choose their own enemy'.(39)

In any event, a test case had already been taken to the Southern Appellate Tribunal in December 1939, in which the Minister appealed against the decision of the London Local Tribunal that a Mr G. Plume, a political objector, should be conditionally registered on the register of conscientious objectors, subject to his remaining in his present employment. The Minister's representative expounded the Ministry's views succinctly, and the chairman warned Plume that he would be on much safer ground if he were to plead pacifism, and not political objection. Fenner Brockway, the chairman of the CBCO, stoutly defended Plume and the principles of political objection by stressing that the decision should not be made on the nature of the conscientiousness, but on the depth and sincerity of the conscientious objection to being on the Military Training Register. He stressed that there was nothing in the Act that 'ruled out' political objection. Unlike in a court of law, the chairman of a Local or Appellate Tribunal did not have to sum up, so that it must be assumed from the failure of Plume's case that the advice of the Minister's representative was heard sympathetically by members of the Tribunal. Some Local Tribunals, however, continued to take Brockway's line and made their judgment on the depth and sincerity of the views held.(40)

Some Tribunals, in fact, refused to recognise a political objection to the war. The CBCO South Wales Regional Board was moved to take a deputation to the Welsh Appellate Tribunal in March 1943 to complain about this.(41) And in 1940 the CBCO reporter at the Fulham Tribunal was amazed when a political objector was told that he could perform non-combatant duties for the duration of the war for the reporter had understood that the ruling of the Fulham Tribunal was that it had no power to grant any exemptions to a political objector, no matter how strong his case might be.(42) This Tribunal had shown little sympathy with political objectors before then. On the other hand, the reporter at the South-Eastern Tribunal, sitting in Southwark, remarked in his general notes that all grounds of conscientious objection were allowed in that Tribunal, even political grounds.(43) An example of the sort of question and answer that would occur when a political objection was raised follows:(44)

Tribunal : Should not a community defend itself?

Applicant : We are not a community but are divided into two parts.

Tribunal : If war was of benefit to the common people, would you support it?

Applicant : It is not of benefit.

Tribunal : Would it be worthwhile to defend an ideal democracy?

Applicant : Socialism is international and if a socialist state started to defend itself, it would no longer be able to retain its socialism.

and then the key question:

Tribunal : Is your objection only against this particular war?

Applicant : No, it is to all wars.

This applicant, who presented his case to the South-Eastern Tribunal in 1940, dealt with these questions in the best possible manner, by not answering hypothetical questions, and by purporting that as a socialist, he could never support a war fought by a so-called socialist state as, on entry into war, it would have ceased to be a socialist state. The Tribunal decided that he had a conscientious objection to military service, and exempted him on condition that he undertook some agricultural or forestry work.

The attitude of the Tribunals to objectors applying for unconditional exemption is best summed up by Joe Brayshaw, an officer of the CBCO, who wrote a memorandum for the CBCO meeting of 5 June 1943, in which he analysed the statistics relating to the granting of unconditional exemptions. Although the Tribunals differed widely in the numbers of this sort of exemption they granted, generally all Tribunals were becoming less and less liable to grant it. The South Midlands, the North Midlands and the North Scotland Tribunals gave no unconditional exemptions at all in 1943. The East Anglian and South-Western Local Tribunals, which had both given quite a high number of unconditional exemptions at the beginning of the war were, by 1943, giving very few. The North Wales Local Tribunal had begun leniently under the Chairmanship of Sir Artemus Jones, but when Judge Samuel took the Chair from September 1941, the number of unconditional exemptions dropped to negligible proportions. The

South-West Scotland, No.2, Local Tribunal was thought to have the worst record for unconditional exemptions of all the Tribunals.(45) The decrease in the numbers of unconditional exemptions given early in the war is quite clear. In 1939, 14 per cent of all objectors were given unconditional exemption; in 1940, 5 per cent, and in 1941, only 2 per cent. As Brayshaw put it: 'unconditional exemption is a dead letter'.(46)

Conditional exemptions

Exemptions on condition that the applicant undertook civilian work under civilian control (called the 'B' decision) were consistently the most common decisions given by Local Tribunals.(47) The work given would normally constitute work of national importance. This, in many ways, seemed to be the ideal solution for many Tribunal members and, indeed, for many conscientious objectors. Tribunal members felt that objectors were not escaping their responsibility to the state and society if they could be ordered to perform work which would contribute towards the nation's efforts to win the war, especially if they were uprooted from their own jobs and moved elsewhere for the duration of the war, as soldiers were. Objectors, meanwhile, who were anxious to show that they were not evading their responsibilities, but who could not reconcile entering the Armed Forces with the dictates of their conscience felt that this was a way of showing their good faith. It was felt, too, by some Tribunal members that, given that men would not join the combatant forces, they would be better employed in useful civilian work, the importance of which was never underestimated, than joining a non-combatant corps, where their contribution would only be of limited value.

The difficulty with this decision lay, therefore, in the sort of work to which the objector should be directed, for the Tribunals had to specify the work. This was a question which worried civil servants in the Ministry of Labour at the beginning of the war, as they felt sure that Tribunals would be looking to the Ministry for guidance on this matter. All kinds of problems were posed by it. At the outset of war, the Government had made a Schedule of Reserved Occupations, which was a list of occupations which it considered to be of vital importance to the conduct of the war, and for the smooth running of civilian society during the war. Many industrial workers, doctors, teachers and other indispensable service workers were included on the list. Usually, an occupation was reserved after a certain age, that is, while men who were young and most suitable for entry into the Armed Forces were called up, even though they were in a Reserved Occupation, men over a certain age, for instance, 30, remained in their jobs, so that the service or industry continued functioning without too much disruption. All men were required to register on the Military Training Register, but only those under the specified age limit were called up. Those who purported to have a conscientious objection to entering the Armed Forces had to register at the same time as the rest of their age group, but their names were provisionally registered on the register of objectors and, in due course, a Local Tribunal decided whether the objection was conscientiously held. Hence arose the problem: should a man, who would normally be called up because of his age, despite the fact that he was in a Reserved Occupation, but who had been instructed by a Tribunal to undertake civilian work, be allowed to remain in his present occupation? If his work was important, as it obviously was if included on the Schedule of Reserved Occupations, it would seem illogical and wasteful to remove a man from this work for which he had been trained. On the other hand, there might be a great deal of public resentment if an objector was allowed to remain in his job, while his colleagues were uprooted and forced to join the Armed Forces. It was a difficult dilemma, but it was decided in the end that objectors should stay in Reserved Occupations and that public resentment would have to be faced, if and when it arose.

Other problems faced the civil servants. There was a danger that certain industries would become 'overcrowded' if Tribunals specified the same sort of work to many objectors. Overcrowding would undoubtedly cause an outcry from the trade unions, but it was in any case a waste of manpower. Tribunals would have to be warned of this possibility. Apart from overcrowding, it was recognised that workers and employers might take exception to the influx of conscientious objectors into their industry, simply because of the views they held.(48)

The civil servants who assumed most responsibility for advising the Tribunals on suitable work for objectors were the officers representing the Minister in Local Tribunals. It was realised that Tribunal members would turn first to the officer for advice on what sort of work to specify, and officers were directed to advise work of national importance; an area of work rather than direction to a specific job. They were told that a test case would be sent to the Appellate Tribunal as soon as possible to determine whether the work specified should in fact be of national importance. Meanwhile, they were not to recommend a job which was already fully supplied with labour, or one that was too skilled, or one that was likely to incur industrial trouble. The work specified for the objector need not be in his own district, but this was thought desirable where possible. Students studying for a job of national importance should be allowed to continue their studies. The Minister, Ernest Brown, requested that the officers should return regular reports of the work specified for his perusal.(49)

At the beginning of the war, the Tribunals seemed to consider that agriculture and forestry were the two industries which could absorb most extra manpower, and therefore specified that work for most objectors who were exempted conditionally. However, there were soon complaints from the Ministry of Agriculture that the industry was not the 'dumping ground' that everyone seemed to think it was and it could not absorb any more new recruits as farm labourers. The public and the press were convinced at this time that there was a shortage in agriculture but, the Ministry pointed out, in certain areas the opposite was true. In September 1939 representatives from the Ministries of Agriculture and Labour held an informal meeting to discuss the position of objectors who had been directed by a Tribunal to obtain work in agriculture. The representatives from the Ministry of Agriculture immediately stated that, if there was room for more labour in the industry, the Women's

Land Army and the unemployed should take preference. The difficulties of the situation were recognised by both sides. Newspaper reports and letters to both Ministries were already making it clear that both farmers and agricultural workers resented objectors being thrust into their industry. A further meeting decided that there might be some possibilities in forestry but that, in any case, the situation ought to be reviewed regularly as it was possible that it might change. Hopes were not high at this stage however.(50)

By January 1940, Tribunal members were alive to the fact that objectors who were directed to land work were finding it increasingly difficult to obtain the work. Judge Davies, of the South-Eastern Tribunal, wrote to the Ministry of Labour asking if Chairmen could be offered some alternative to agriculture and forestry. The reply was sympathetic, but offered no alternative.(51) In fact, as the war progressed, there developed a shortage of agricultural workers, and the situation was, therefore, eased. Even in 1939 the manpower situation had not been as simple as either the press or the Ministry of Agriculture had depicted. There was neither a huge shortage nor a huge surplus in the industry, but rather a very uneven distribution of manpower. While both surpluses and shortages existed in the industry, the situation was bound to be confusing.(52)

Meanwhile Tribunals made serious efforts to direct objectors away from agriculture and forestry, until the situation became clearer, and into other areas of employment. Before the 1941 Act, Civil Defence was not a compulsory duty and, if the objector seemed willing, full-time Civil Defence appeared an altogether more suitable alternative. Local Government services, such as the fire brigade and ambulance services, offered possibilities, but the decisions of some local authorities not to employ conscientious objectors prevented many of them from obtaining work in these areas. Many Tribunals eventually came to the conclusion that objectors, who were already usefully employed, although not necessarily engaged in work of the utmost national importance, were best left to continue in their present employment. This decision was especially desirable if the objector was found to be undertaking spare-time Civil Defence work already. It was a compromise between original idealistic aims and immediate practical necessities, but it appeared to be the most satisfactory solution to the problem of specifying work for conditionally exempted objectors.

Exemptions from combatant duties in the Armed Forces

The third decision, or 'C' decision, that the Tribunals could give, was to register the objector as liable to be called up for non-combatant duties only.(53) Most who remembered the First World War, and this included almost all of the Tribunal members, also remembered the sterling work conscientious objectors had performed in the Royal Army Medical Corps (hereafter RAMC). This, Tribunal members believed, could be repeated in the Second War. So when it became clear that an applicant objected solely to combatant duties in the Armed Forces, Tribunals directed him to non-combatant duties in the Armed Forces with a recommendation that he should be allowed to join the RAMC. It had to be a recommendation only, since Tribunals had no power to say which unit the objector should join; that was a matter for the Armed Forces. But until mid-1940 most of those who were recommended for the RAMC did manage to serve in it. In fact no corps, except for the Royal Army Chaplain's Corps, was officially non-combatant. In theory, at least, a member of the RAMC or the Royal Army Dental Corps or any other corps which did not usually engage in combat, could be ordered to take up arms, and there was some confusion in 1940 about exactly how possible or likely a situation of that kind could develop. However, all was resolved in April 1940, when a Non-Combatant Corps was formed expressly for the purpose of receiving objectors into the Army.

The eagerness of Tribunal members to recommend applicants to the RAMC is revealed in all the Tribunal records. A CBCO reporter in the South-Eastern Tribunal noted in November 1939 that one of the members, Sir Reginald Kennedy-Cox, had been a member of the RAMC in the last war. 'This may account for the readiness to accept any applicant who will undertake RAMC work with little or no question.' (54) In February 1940, when the first doubts arose among pacifist organisations about the combatant or non-combatant nature of the RAMC, a reporter in the same Tribunal noted that pressure was being brought to bear on objectors, with one of the Tribunal members, Dr Senter, in his efforts to persuade objectors to join the Armed Forces as a non-combatant in the RAMC, frequently saying, 'You know, don't you, that you will not be required to bear arms.'(55) The Fulham Tribunal was also well known for habitually pressing objectors to join the RAMC.(56)

Pacifist organisations only really became worried about this keenness on behalf of the Tribunal members to recruit for the RAMC when two factors became clear. The first was the suspicion that, despite their Tribunal hearing, objectors might have to take up arms, since the RAMC was not an officially non-combatant corps. However, this difficulty was resolved in March 1940 when the War Office made it clear that a man performing non-combatant duties in the Armed Services would never be required to drill with or handle lethal weapons.(57) The second, and more serious, factor as increasing numbers of objectors entered the RAMC, was that the corps was becoming overmanned and only required skilled recruits. Before the creation of the Non-Combatant Corps in mid-1940, conscientious objectors who had agreed to perform non-combatant work solely on the understanding that they would be joining the RAMC were being told, once in the Armed Forces, that there was no room for them in that Corps. Whether Tribunal members knew of this overcrowding is debatable. Some chairmen, such as Judge Davies of the South-Eastern Tribunal, definitely warned applicants that, although he might recommend them for the RAMC, he could not ensure that the recommendation would be fulfilled. But Sir Gerald Hurst, the chairman succeeding Judge Davies, apparently gave no such warning, much to the reporter's consternation.(58) Indeed some Tribunals were wording their decisions that the applicant 'shall be liable for military service only in RAMC duties.' Civil servants advised the Minister to appeal against one decision like this in 1940, and the Appellate Tribunal duly ruled that the words, 'only in RAMC duties' should be substituted with the words, 'only in non-combatant duties', on the grounds that the original order was not in accordance with sub-section 6 of section J of the National Service (Armed Forces) Act, 1939.(59) This 'test' case provided a precedent to which Local Tribunals could refer and thereafter they adhered to the principles of ordering non-combatant duties with no more than a reccommendation for RAMC work.

The other major step in the area of non-combatant decisions was that in 1941, when the National Service Act (Number 2) was introduced, applicants who were given the 'C' decision no longer had their names removed from the register of conscientious objectors. The decision to direct an objector to non-combatant work was now treated as a form of conditional exemption from military service in the same manner as the 'B' decision. It seems a small matter, but to the objector it was important, for their exact status as conscientious objectors, even when they were part of the Armed Forces, was now recognised. With the formation of the Non-Combatant Corps, the 'C' decisions could be given with far less doubt and suspicion on the part of both objectors and Tribunal members than was possible before.(60)

Failed applications

If the Tribunal was not satisfied that the application was genuine, it could direct that the applicant's name should be removed from the register of conscientious objectors and that he should be liable for military service, the 'D' decision.(61) How Tribunals reached their conclusions on whether or not an application was genuine, and indeed in which of the other categories it should be placed if it was thought to be genuine, will be discussed in the next chapter, but one way of finding whether the applicants who had been rejected were satisfied with the decision on their individual cases is to look at the number of appeals there were against the 'D' decisions in the Local Tribunals.

Of the 18,495 men and women whose applications had been rejected by the Local Tribunals up to the end of 1948, the period which the Ministry of Labour statistics cover, 10,878 felt aggrieved and appealed to the Appellate Tribunal. Of the rest who did not appeal, some accepted their fate and entered the Armed Forces: others, believing perhaps that an appeal would bring the same result, refused the medical examination for entry into the Armed Forces, and were imprisoned or fined. Of the 10,898 protesting applicants, 5,852 failed to get their 'D' decision varied at the Appellate Tribunal; altogether 1,891 served terms of imprisonment for refusal to submit to medical examination for the Armed Forces. Therefore the majority of those who were given the 'D' decision at the Local Tribunal, and,for those who persisted, again at the Appellate Tribunal, entered the Armed Forces without further objection.(62)

Non-compliance with a condition

The only other decision that the Tribunals were required to make by law concerned the fate of those who had failed to comply with the condition on which they had originally been exempted. The 1939 National Service Act laid down that 'if on the information of any person' a Local Tribunal was satisfied that a person had not complied with his condition, it should report to the Minister, who would require the person to make a fresh application to the Local Tribunal. The Tribunal could then rehear the case and decide what should be done. If he failed to make a fresh application, the Minister could remove his name from the register of conscientious objectors, and register him as a person liable to be called up, but to be employed in non-combatant duties only.(63)

The 1941 National Service Act, providing for the call up of persons for Civil Defence duties, also made some amendments to the 1939, or 'principal' Act, repealing the section relating to breach of condition of registration as a conscientious objector. The new Act introduced a novel term into this subject, that of 'reasonable' excuse'. If the Minister considered that a person had reasonable excuse for his failure to comply with a condition, he referred the person's case to the Local Tribunal. The Tribunal, if it agreed with the Minister, could either make no order in the matter, or order that the person should be registered without conditions on the register of conscientious objectors, or order that the condition should be varied or another substituted. Both objector and Minister could appeal against this decision. If the Tribunal found that the applicant had no reasonable excuse for his failure to comply, he had committed an offence under the principal Act, and was liable to a maximum of two years' imprisonment or a maximum fine of £100, or both.(64)

The more elaborate provisions in the second Act helped a great deal to solve the problems of those who had been awarded the 'B' decision and of those whose job it was to see that conditions were complied with. It showed that the Ministry of Labour understood that it was possible that a person, despite every effort, might not have been able to obtain the work the Tribunal had specified, and ought therefore be allowed to look for different work, or to return to his original occupation. Under the principal Act some conditionally exempted objectors had been unemployed for months on end. As a Ministry of Labour official pointed out to the 'Yorkshire Post', 'So long as a man was making efforts to obtain work ordered by the tribunal, and maintained contact with the local labour exchange, the authorities must accept his efforts as reasonable.'(65) Now the Minister could, without any implied criticism of the objector at all, refer the case back to the Tribunal so that the waste of manpower could be halted. If, on the other hand, the Minister was satisfied that the objector had made no reasonable effort to obtain work, the objector could be prosecuted in the ordinary way, and tried in a court of law for his offence. No prosecutions, however, could be brought without the Minister's consent.(66)

One other possible source of confusion remained. If an objector had reasonable excuse for his failure to obtain the specified work, the Local Tribunal could vary the condition on which he was exempted, or substitute a new one for it. The 1941 Act, as has been explained, made non-combatant service a form of conditional exemption. It was, therefore, possible that certain Tribunals might change the condition of exemption to performing non-combatant service, even though the applicant had originally been exempted from serving in the Armed Forces. The CBCO became especially worried about this when one of its officers, Robert Polland, who had represented an applicant at an Appellate Tribunal, reported that the members of the Tribunal had appeared sympathetic to the argument of the Minister's representative that, as non-combatant service was now a form of conditional exemption, it could therefore be given as an alternative when an original condition had not been complied with.(67) But unless an objector had changed his mind on whether or not he felt he could serve in the Armed Forces as a non-combatant (and at his original hearing it was obvious that he had successfully pleaded that he could not), then it was unlikely that the Tribunals would use this, albeit legal, manoeuvre to force a man into the Armed Forces as a non-combatant against his will. Certainly there are no further records that this ever occurred; it is not mentioned again in the CBCO records.

After the 1941 Act was passed, and until the end of 1948, 6,564 cases were heard in Local Tribunals from objectors who failed to comply with their original condition, 5,443 of which resulted in the condition of registration being varied. 738 were registered unconditionally and 104 cases had no fresh order made. Despite their appearance before the Tribunal, 90 were found to have actually complied with the condition and 189 were found to have had no reasonable excuse and they were prosecuted in a civil court. Some 485 people appealed against these decisions, 353 of which had their condition varied.(68) The Tribunals, therefore, played a vital and useful role in the 'follow-up' procedure of conscientious objectors conditionally exempted from military service.

This chapter has dealt with the constitution, functions and formal procedures of the Local and Appellate Tribunals, and with the possible decisions which the Local Tribunals could make. It now remains to study on what criteria Tribunal members were selected, on what grounds they based their decisions, to study those who were presenting their case before Tribunal members, the applicants themselves, and lastly to assess the work of the Conscientious Objectors Tribunals in the Second World War.

Chapter 3

TRIBUNALS IN ACTION: their work and an assessment of it

THE MEMBERS

One of the main criticisms of the conscientious objectors' Tribunals in the First World War had been of their membership. They were supposed to be composed of members of 'impartial and balanced judgement' (1) but in fact very little machinery was put into operation to secure this objective. Since the First World War Tribunals were a development of the Derby recruiting-scheme Tribunals, the membership of the two bodies tended to remain the same, so that members who were virtually recruiting officers were being asked to transform themselves into judges of pleas for exemption from the Armed Services, a change of roles difficult, although not impossible, to make.

Both John Rae in 'Conscience and Politics' (2) and John Hughes in his thesis, The Legal Implications of Conscientious Objection (3) go a long way towards a condemnation of the membership of the First World War Tribunals on this and other grounds. Rae maintains that members strongly supported the ethic of war, except for a small number of Quakers who were appointed onto the Tribunals; that the War Office and the Local Government Board, who had joint responsibility for the Tribunals, exercised very little control over them, although Walter Long, President of the Local Government Board, appointed personally the members of the Appeal Tribunals; and that the close relationship between members of the Tribunals and the military effectively made members, in spirit at least, continue to be part of the War Office team. Lastly, he mentions that Tribunals were invariably composed solely of the membership of the local Town or County Councils. Hughs complains that the membership rarely had any legal experience and was often composed of 'local worthies'. Representatives of Trade Unions would often find it difficult to attend the Tribunals during the day, and no fees were paid, again discouraging the appearance of Labour or Trade Union representatives. He mentions that few women were appointed members of the Tribunals. Finally, he stresses the lack of detailed rules for the members of the Tribunals to follow, and points out that the Local Government Board, while giving advice, failed to make its advice compulsory.

All these criticisms raise one fundamental question, that is, what sort of person does in fact make for the best judge of conscience, if indeed, such a task is possible at all? Both authors assume that local councillors were not necessarily the best men for the job. Hughes obviously feels that men and women with some sort of legal experience were essential, although, as will be seen, one MP in 1939 thought that men of legal standing were the most unsuitable group of people to act as judges of conscience. The fact that most members of the Tribunals supported the fighting of the First World War and wished to see the maximum number of men recruited does not of itself mean that they would be incapable of judging a conscientious objection to it. The logical step would be to employ only conscientious objectors as members of the Tribunals, but this assumes that only conscientious objectors can judge conscience: surely a contentious viewpoint. Finding enough people who were utterly indifferent to the issue would be well-nigh impossible, and, in any case, would not alter the fact that the attitude of a man to war does not of necessity cloud his judgment of conscience. It was indeed a pity that the men who administered the Derby Scheme should carry on to pass judgment on those claiming exemption from compulsory recruitment but in theory it was not impossible for men and women to alter their aims and objectives as Tribunal members. Neither author disagrees that there should have been some representation of the Trade Unions in the membership, although why they think this is worthwhile they do not explain. They both criticise the lack of centralised control over the selection of the members of the Local Tribunals, but again fail to state on what criteria any selection by Central Government should have been based.

Perhaps the fundamental question is unanswerable. At the beginning of the Second World War the Government promised in the National Service (Armed Forces) Act, 1939, that the Minister of Labour and National Service, in appointing members of the Tribunals, 'shall have regard to the necessity of selecting impartial persons' and that 'The Chairman shall be a county court judge or, in the case of a local tribunal for a district in Scotland, a sheriff or sheriff-substitute.' In the Appellate Tribunals, the chairman would be nominated by the Lord Chancellor or, in Scotland, by the Lord President of the Court of Session.(4) The government of the day, therefore, had opted for the presence of a legal authority on each Tribunal, indeed in the Chair of each Tribunal, to ensure impartiality and good judgment. However, when the Bill was debated in the House of Commons, C.C. Poole MP (Labour) took exception to the choice of county-court judges as Chairmen of the Tribunals.(5)

> You are not trying a man for any crime when he appeals to a Local Tribunal to be exempt.... I cannot conceive that you will have the most sympathetic consideration, as you might reasonably expect to have, by having a county court judge in the Chair.... After all, it is not desired in these cases that you should have someone who can weigh up evidence. You want someone who has a great and deep understanding of human nature. You want someone who will be sympathetic and who will endeavour to put himself in the place of the person who is appearing, and not someone who calmly and dispassionately weighs up the pros and cons. Many of these men will be working class men who will not be able to state their cases adequately, and if they are to be left to the tender mercies of barristers and solicitors of not less than ten years

> standing, they will have a very poor chance of establishing their case.

However, Poole's voice was a lone one. While many MPs were concerned with the choice of chairmen and members, no others questioned the policy of appointing county-court judges as chairmen. Indeed, there is no evidence to suggest that any significant body of opinion objected to this choice. The CBCO, while criticising many individual chairmen, never raised the question of the desirability or otherwise of having judges as chairmen at any of their meetings. The reasoning was that, of all public figures, judges were most likely to be impartial because of the very nature of their profession. Impartiality, then, was the key criterion for choosing a chairman, not a facility for judging conscience. Sir Arnold T. Wilson MP (Conservative), speaking in a debate on the Military Training Bill in 1939, thought the task impossible.(6)

> The dictates of conscience are not admitted in any court of law, as, for example, in the case of the so-called 'mercy murders'. In the words of a great judge in the reign of Henry VIII: 'The mind of man is not triable.' We are endeavouring in Clause 3 of the Bill to try the mind of man, and I foresee, as in the last War, great difficulties.

If the mind of man was 'untriable', the Government could at least attempt to put on the Tribunals persons who would be as impartial as possible in endeavouring to judge the sincerity of a man's views, which, in the end, was all that Tribunals could set out to discover.

T. Edmund Harvey MP (Independent Progressive), a conscientious objector from the last war, certainly believed that the absence of lawyers on the First World War Tribunals had been a serious omission. While he thought that, 'There is no machinery which the House or any other House can set up for judging the consciences of men that can be satisfactory,' he considered that the provisions made for conscientious objectors in the Bill of 1939 were far better than those made in 1916:(7)

> In particular, the judicial character of the tribunals is a matter of the greatest importance and value. I can remember how a bishop - not a pacifist bishop - wrote to 'The Times' to protest about the behaviour of the tribunals and said that they reminded him of the conduct of Mr. Justice Hategood in Bunyan's 'Pilgrim's Progress'. I hope nothing of that sort will have to be said of the tribunals that are to be set up under this Bill when it becomes an Act.

George Lansbury, who had lost the leadership of his party on the issue of pacifism, was more critical:(8)

> I do not think that old men like me ought to be put on these tribunals to weigh up the judgement and the conscience of much younger men, but persons much younger and persons with an appreciation of what conscience means.

It is fair to say, however, that MPs were less concerned with exactly who sat on the Tribunals than that their nature should be entirely civilian. Many objectors and impartial observers of Tribunals in the last war had considered that they amounted to little more than recruiting centres for the Armed Forces, a belief partly substantiated by the fact that at many hearings a military representative was present, acting very much like a prosecution counsel.

J. McGovern MP (Independent Labour Party) voiced the doubts and suspicions of many MPs when he said in the same debate: (9)

> He [the Minister of Labour and National Service] gave an assurance earlier that the personnel of the tribunals would not constitute of military people, but I would point out these words in the proposed addition to the Clause: 'The Minister or any person authorised by him shall be entitled to be heard on any application to a tribunal.' Can the right hon. Gentleman give us an assurance that those who represent the Minister at these tribunals will not be military people?

Even Lloyd George, whose own Government had severed the connection between the military and recruiting, wanted to be sure that the Minister would exclude 'pukka soldiers' from the Tribunals. (10) The Minister, Ernest Brown, repeatedly assured members that the Tribunals would be entirely civilian in character, although he did point out that it would be undesirable to disallow retired military persons from acting as members of the Tribunals or representing the Minister. He stressed that the whole procedure for setting up and administering the Tribunals was his responsibility, and that it had nothing to do with the War Office as in the First World War: 'the Minister is responsible both for appointing the tribunal and for seeing that its decisions are carried out.' (11) By February 1940, the Minister was assuring members of a Committee of the House: (12)

> I have really tried to face up to my duty .... in selecting the members of the tribunals .... the Noble Lady [Viscountess Astor] hoped that great care would be taken in this matter. I can assure her that great care has been taken. Members of long experience, I am glad to say, expressed their appreciation several times of the difficulties of the problem. It was not always easy to get the tribunals formed. One of the best members of a particular tribunal replied to me, when I asked him to serve: 'It is a public duty and it is so disagreeable that I cannot refuse to do it.'

He went on to say that he had taken particular care in securing the impartiality of Tribunal members.

The Minister, on his own admission, found some difficulty in finding suitable members, and it is interesting to see which people he finally appointed onto the Tribunals. In April 1940, T. Edmund Harvey MP asked the Minister for the names of all the chairmen and members of the Local and Appellate Tribunals. The Minister complied with the request and a list appears in Hansard. (13)

In 1940 there were only two Appellate Tribunals, one for England and Wales, and one for Scotland. The chairmen of these were the Oxford historian, the Right Hon. H.A.L. Fisher, and the Right Hon. Lord Elphinstone, formerly Governor of the Bank of Scotland. The Local Tribunals consisted of a chairman and four members. The chairman, as has been explained, was always a county-court judge, or, in Scotland, a sheriff or a sheriff-substitute. The 1939 Act also stated that of the four members 'not less than one should be appointed by the Minister after consultation with organisations representative of workers.' (14) The idea behind this was presumably that some representation on the Tribunals of the 'common man' was desirable; in any case, examples of men who sat on the Tribunals in this capacity are, from the London Local Tribunal, A.B. Swales, a former vice-chairman of the General Council of the TUC, and from the North-Western Tribunal, A. Roberts, an ex-chairman of the TUC.

Apart from the prerequisite of impartiality, the Act left the choice of the rest of the members completely in the hands of the Minister. Interestingly, there was almost invariably an academic of some standing on each Tribunal. On the South-Eastern Tribunal was Dr G. Senter, principal of Birkbeck College, University of London. J.H. Clapham, who was later knighted, professor of Economic History at Cambridge University, was a member of the East Anglian Tribunal. JPs from various professions formed another group widely represented in the membership of the Tribunals. Sir Reginald Kennedy-Cox, a playwright and author, a governor of Malvern College, and the founder of the Dockland Settlements, which gave relief to dockers undergoing hardship, was one of the members of the South-Eastern Tribunal. Charles Aveling and Sir Miles Ewart Mitchell were both members of the North-Western Local Tribunal; the former had been Mayor of Southport, the latter Lord Mayor of Manchester.

Other professions employed included that of law, in the form of solicitors and barristers, medicine (Peter Campbell of the Northern Scottish Tribunal was a surgeon), chartered accountancy, insurance, and journalism which was represented by Sir Robert Bruce, who had been editor of the 'Glasgow Herald' until 1936, and who sat on the South-Western Scottish Tribunal. The average age of chairmen and members was about 65, although there were at least three members who were over the age of 80 in 1940. One of the youngest men must have been A. Roberts of the North-Western Tribunal, who was only 42 in 1940. Despite Lansbury's wish that the Tribunals should be composed of young men, the Minister obviously believed that age and experience were key factors in securing impartial Tribunals.(15)

Since the members of the Tribunals were chosen on the basis that they would provide an impartial hearing for applicants, an assessment of their impartiality is required. In order to reach a decision on each applicant, the members asked questions of the applicants, and based their decisions on the answers given to them. A study of the sort of question that was put to applicants, and members' reactions to the answers is, therefore, necessary before an assessment of impartiality can be made.

If an applicant objected on religious grounds, members frequently asked whether the applicant belonged to any religious group or church. If he did, they would then ask for how long he had been a member and on which specific religious tenets he based his beliefs. If he belonged to the Church of England or to the Roman Catholic Church, members would often point out that the leaders of those Churches supported the fighting of the war. If he belonged to a Church which had a long history of pacifist doctrine, as for instance the Christadelphians, members put more emphasis on the length of membership of the church than on what the church believed, since it was obvious that if an individual was a sincere and obedient member of the church, he would accept implicitly the teaching of that church. But some Tribunal members tried to show that those belonging to such churches did not understand the basic tenets of their beliefs, and, as such, could not really have a truly conscientious objection. Members' suspicions that Christadelphians had limited understanding of their church's teaching were partly substantiated by the fact that many applicants of that faith appeared in the Tribunal with standard statements from which they read.

Judge Davis of the South-Eastern Tribunal mentioned during a hearing his dislike of that kind of 'stereotyped document'.(16)

When the applicant maintained that war and fighting were contrary to the teaching of the Bible, members would ask to which passages specifically he referred. The Commandment 'Thou shalt not kill' was frequently discussed in Tribunals. Members offered different interpretations of the Commandment to test whether the applicant had really considered the problem. In the London Tribunal, Sir James Baillie commented that, 'Thou shalt not kill was a spiritual and not a physical law' (17) and 'I suppose you know that [the Commandment] only refers to murder.'(18) Occasionally, arguments over the interpretation of the Bible became so complicated that, at the London Tribunal, a CBCO reporter commented that theologians were needed to argue some of the Christian tenets at issue.(19) Applicants who based their objections on religious grounds very often merely objected to the act of killing, but did not conscientiously object to war as such. Those who objected to the whole concept of war, and who did so because of their religious beliefs, were very closely questioned by Tribunal members. To understand why a Christian refused to join the Royal Army Medical Corps, one applicant was asked 'Would He particularise between soldiers and civilians in our streets?'(2

No case was typical but here follows an example of the sort of dialogue which occurred when a religious objector came before the Local Tribunal in London. A stock-taker and gauger appeared before the Tribunal in 1940 with a written statement saying that he was a Christian pacifist, and that he believed that war was contrary to the teaching of Jesus Christ. He took literally the Commandment 'Love thy neighbour', and had determined to face the threat of evil with charity and love. Moreover, he felt that it was a Christian duty to overcome evil with good, for 'we should do as we would be done by.' He added that he was a member of the Church of England, and that he was willing to do ARP work and had already spent his spare time working with that organisation. The Chairman asked if criminals should be 'shut up' or if they should be let out of prison in the name of love. The applicant replied that he did not think criminals should be made prisoners but he appeared confused and bothered by the cross-questioning. A member of the Tribunal commented 'If you had a conscience it would tell you where you stood.' The applicant said it was wrong to bear arms and when a member mentioned the Christian crusades, the applicant said he thought many young men who were Christians had no scruples. A Congregational Minister appeared before the Tribunal to speak on the applicant's behalf. He said that he had known the applicant for six months, and that he believed him to be dissatisfied with the Church of England. He had had many talks with him, and they both believed that the need was for Christian missions, not for 'bayonets and poison gas'. He mentioned Dr Howard Somerville and his medical missions. The witness said that the applicant was sincere, but found it difficult to express his views: however, the Christian conviction was genuine. The Tribunal decided that it was not satisfied that there was a conscientious objection, and the name was removed from the provisional register of conscientious objectors and placed on the military-service register. (21) Inadequate and confused answers to the members' questions probably accounted for this decision.

Applicants who based their objections on moral beliefs were obviously questioned rather differently from those who objected on religious grounds. The reporter at the South-Eastern Tribunal recorded a number of 'stock questions' addressed to this type of applicant, that is, questions regularly asked, and therefore presumably questions which members felt drew answers from the applicants showing whether or not they had a genuine conscientious objection. The questions included

> Don't you realise that many of your old school/work friends are risking their lives to defend you?
>
> Won't you be ashamed to look them in the face if you shirk your duty now?
>
> Would you defend yourself from attack?
>
> Would you not defend your own wife/parents from the ruthless invader?
>
> Was Poland/Norway committing a sinful act in resisting the ruthless invader?
>
> Would England be committing a sinful act in resisting the ruthless invader?

The reporter commented cynically that if the applicant answered 'No' to the first three questions, he was believed by the members to be either a liar or a fool. If he answered 'Yes', his case was undermined. To answer 'No' to the last two questions would be to take away the grounds for the conscientious objection, but the answer 'Yes' would reveal signs of insanity!(22) The reporter's reactions to these questions, while obviously a result of his own bias against the Tribunal members and for the applicants, does show how remarkably difficult it must have been to ask questions which did not put the applicant in an invidious position but, on the other hand, did test his sincerity and depth of feeling. Another example of this sort of questioning, the appeal to the supposedly better sense of the applicant, is the question 'If you saw a train accident you'd go and help, wouldn't you?'(23) This is exactly the same sort of question as that addressed to religious applicants asking if God would particularise between a soldier or a civilian lying wounded in a street: the one appealing to the purely moral sense, the other, to a religious sense. The questions may seem unfair and out of context, but it is not necessarily true that members would really consider the case undermined if the applicant replied that he would help soldiers and civilians, or that he would help the victims of a train crash. The Tribunals were looking for the applicant who could show in what way his refusal to join a non-combatant corps differed from a refusal to perform the other tasks. If he had thought deeply about the question of his refusal, members believed that an applicant would be able to make some sort of distinction between the two. If he could find no answer, then the members would assume that the applicant had given little or no thought to his refusal, in other words, he could not really have consulted his conscience, and, therefore, could not be said to object conscientiously. This reasoning was perfectly fair assuming that the applicant had the intellectual ability to think in this way, and, even more importantly, that he was sufficiently articulate to express clearly and convincingly for what reasons he had reached his conclusion. Otherwise a situation rapidly might have developed where only the intelligent and the articulate were able to object conscientiously to combatant or non-combatant service. The statement and the answers to questions asked of a witness were, therefore, crucially important. A reporter in the London Local Tribunal wrote that the production of a witness or a letter supporting the application was very often a deciding factor in the granting of exemptions.(24) On the reasoning propounded, this procedure in Tribunals was highly desirable.

In attempting to judge the impartiality of the Tribunals it is easy to fall into the trap of noting only those occasions, and they were not uncommon, when Tribunal members were patently not impartial. Comments like that made by Judge Richardson at the Newcastle Tribunal, 'It is a pity we cannot put you people [conscientious objectors] on a desert island so that you could all enjoy yourselves' (25) clearly did nothing to create a judicial air of impartiality in that Tribunal. Petty bickering between applicants and members was not unknown. At the North-Eastern Tribunal, an applicant was asked if he was paying income tax. This was a widely used method of eliciting from applicants who were applying for unconditional exemption, whether they had fully considered that, by taking part in society at all, they were in some way assisting the war effort. The applicant replied that he realised the question was a common trap for pacifists. A member of the Tribunal said that there was no trap. It was the applicant that was talking 'claptrap'.

Applicant : I didn't expect personal insults when I came here.

Chairman : You are just a prickly fool. You are not worth insulting. 'Claptrap' is not an insult but a statement of fact.

A member observed that the applicant had insulted the Tribunal in the first place by saying that it was trying to trap him.(26) Remarks like those of Sheriff Brown, Chairman of the South-Eastern Scottish Tribunal, 'We will call you the new contemptibles'(27), or of Judge Frankland at the Manchester Tribunal, 'The room will be pleasanter when you have left it'(28), or of Judge Burgis at the North-Western Tribunal, when he told an applicant who was employed on gardening and kitchen duties in a Quaker hostel, that he was doing 'women's work', (29) were unnecessary and irresponsible outbursts provoking bitterness, resentment and mutual distrust between members and applicants.

There is no doubt that some Tribunals were prone to this kind of behaviour, but it would be surprising if the Minister had been able to make the best possible choice of members for each Tribunal. There were bound to be some wrong or doubtful choices made. However, it was quite clear when a particular Tribunal was becoming vindictive, bullying or in any way losing an impartial, judicial atmosphere. The local press was invariably present at hearings, and quickly made public any sufficiently sensational exchanges. Reporters and observers from organisations representing objectors promptly complained at any untoward activity, and there were a number of MPs who took an interest in events at Tribunals and were ready to act if they felt it necessary. If, then, a study is made of local newspaper reports, the minutes of the conscientious-objector organisations, and of parliamentary reports, it is at once obvious that, while some Tribunals were criticised frequently, a much larger number were never mentioned critically. One can fairly assume that these Tribunals were being as impartial as the Act had intended them to be. What transcripts are left of the Tribunal hearings bear out this conclusion.

THE APPLICANTS

In 1942 Judge Wethered of the South-Western Tribunal wrote a memorandum, which was not published, describing the objections raised by conscientious objectors in his Tribunal. Wethered heard 4,056 cases up to 7 March 1942, and of these, 71 per cent based their objections on religious grounds and the rest on moral or political grounds. In some ways, this description of the types and numbers of objectors appearing before Wethered is untypical. For instance, Wethered felt that he saw proportionately more Methodists than other Tribunals because there happened to be a preponderance of Methodists in that part of the country. The sorts of people living in a largely rural and economically prosperous area would, to some extent, determine the sort of objectors appearing before him. Additionally, the physical threat of war was not as intense as it was, say, in the South-east and in London, and in the big industrial towns in the Midlands and in the North, where air raids were a depressingly common occurrence. However, despite these points, Wethered's report sheds some light on the numerical proportions and the types of objections made in Tribunals in the Second World War.

Wethered separated the religious objectors into three groups: firstly, those belonging to bodies whose association was based on Second Adventist expectations; secondly, those belonging to Protestant churches; thirdly, Roman Catholics. A belief in the Second Advent of Christ is the basic tenet of the Christadelphians, the Plymouth Brethren and the International Bible Students Association or Jehovah's Witnesses. Of the total number of objectors whom Wethered saw, 170 claimed to base their beliefs on the tenets of the Christadelphians, whose objection to war had been recognised since the American Civil War. Wethered noted that very few of them objected to performing work of national importance, although some felt that they could not work for an 'earthly ruler'. These objectors, however, formed a break-away group from the main body of Christadelphians. Some 439 members of the Plymouth Brethren appeared before Wethered, and he found them easier in their attitude towards military service in that they would undertake non-combatant work in most cases. The Brethren appeared 'grateful to the Government and people of this country for their hospitality.' Jehovah's Witnesses, however, were much more uncompromising in their attitude. They were completely neutral about the war, and accounted for it by saying that it was 'God's wrath' which made nations fight each other; 155 of them appeared before Wethered. The Judge had a certain respect for Second

Adventists since none of them claimed a total right of citizenship, and were therefore logical in their objection to fighting for a country to which they felt no particular loyalty.

Of the Protestant adherents, 662 Methodists appeared before Wethered. He found their attitude equivocal:

> Their position is the result of simple Bible teaching operating on a mental background of almost complete ignorance of the external world, outside a very small circle of home, friends, work and Chapel.

The Baptists, of whom there were 187, and the Congregationalists, of whom there were 143, Wethered found similar in outlook and beliefs. Altogether, 531 applicants stated that their religion was Church of England and purported to base their objections on the teaching of that church, although Wethered discovered that about half of these were not communicant nor active members of the church. They frequently used the phrase 'the way of the cross' and emphasised that Christ taught how evil must never be resisted by the use of force. There had to be an absolute submission to wrong-doing, in the faith that, under the providence of God, good would certainly prevail. Wethered found these views irritating:

> the religious convictions of the applicants are resolutely held without any sense of the necessity of testing theoretical beliefs in the light of practical experience. This essentially unscientific attitude to Truth is indeed characteristic of the whole body of Conscientious Objectors whom we have met.

Wethered also found the beliefs of the Quakers contradictory and illogical. He described the faith, showing how the Quakers believe that the natural conscience is variable and uncertain, but that there is something vastly superior to conscience in each man's soul, 'the Light within'. This is a 'super-added' natural gift which cannot be corrupted or influenced: 'Conscience is thus compared to a lantern and the light within in the candle which illuminated it.' But Wethered argued that if the natural conscience was admitted to be capable of incorrect decisions, how could it be said that the 'awful voice of conscience' or 'the final arbiter' was speaking? Many who appeared before Wethered, and there were 302 of them, were absolutists, a position which Wethered found particularly difficult to understand, since Quakers claimed full rights to citizenship and yet refused to defend their country, whatever atrocities might be performed. However, Wethered respected and admired the Quakers' social activities, for instance, the Friends' Ambulance Unit. Other Protestant sects which Wethered encountered were the Presbyterians, of whom 12 made applications to his Tribunal, and the Unitarians of whom there were only 9. Some 18 Jews appeared before Wethered, but this was not on account of their religion, but due to personal moral grounds. Just 9 objectors of Indian origin appeared and they usually belonged to religious groups in India which had rejected the concept of war.

The third grouping which Wethered defined, the Roman Catholics, were represented by just 64 applications. The Catholic concept of a 'just war' Wethered thought logical, although he felt it impossible to say what constituted a 'just' war. While the Cardinals in Britain had instructed Catholics that the war was just, the Pope had not spoken on the matter.

Wethered dealt with over 1,000 objections on moral grounds. These applicants always said that their objection had no religious foundation, but it was often clear to Wethered that religious beliefs influenced their thinking. He thought that these objectors generally read widely and that some were cynical. He predicted that some would revise their opinions as the war progressed, as there had 'already been a number of withdrawals'. Typical remarks from this kind of objector were:

All wars are futile.
War does not solve anything.
War breeds war.
There will be a Peace Conference at the end of the war-why not now?
The British Government declared war on Germany.
The Treaty of Versailles is to blame.
We have acquired our Empire by conquest.
The German people do not support Hitler, why should we kill them?
The German people are no worse than we are.

Wethered was worried that many of the applicants were in, or were training to be in, the teaching profession: 'Many of them are not suitable persons to be entrusted with the instruction of the Nation's youth.' He commented that a lot of the moral objectors had 'stupid views' but views which were most sincerely held: 'They are the creation of pacifist teachers and workers during the last 20 years.' Political objectors he found very difficult to deal with, especially as, of all the objectors, they frequently held their views with the most tenacity, and most fervently.(30)

One popular conception of conscientious objectors is that they were all left-wing middle-class intellectuals: teachers, writers and academics. It is impossible to make an accurate statistical survey of the occupations and backgrounds of all objectors in the Second World War; there are just not enough records left. However, if the occupations of applicants, appearing before a Tribunal in one day taken at random with no evidence to show that it was in any way exceptional, are examined, it suggests that the popular conception is wrong. In the London Local Tribunal, on 19 December 1939, the applicants who appeared were composed of two shipping clerks, a labourer, a sales manager, a linotype operator, an assembler and distributor of office equipment, a telegraph fitter, a baker, an insurance clerk, a paint sprayer, a Sainsbury's salesman, a library assistant, an architect's draftsman, a civil-service clerk, a shop assistant, a snack-bar attendant, a solicitor's clerk, a cafe manager and a clerk in a trading company.(31) If anything, this shows a preponderance of 'white-collar' workers. However, no definite conclusions should be drawn from such a small sample. A better picture can be depicted by the study of figures over a longer period. In a three-month period the following occupations were represented at the England and Wales Appellate Tribunal: 117 artisans, 109 clerks, 66 students, 62 shop workers, 57 unemployed, 45 professional workers, 37 civil servants, 25 local Government workers, 22 labourers, 22 transport workers, 11 agricultural workers, 8 chemical workers, 7 teachers, 5 self-employed and 2 policemen.(32) These figures were collected by a reporter for the CBCO and it is unclear in some cases exactly what he meant by the group classifications. For instance, the word 'artisan' could include numerous occupations. What is meant by 'professional workers' is also unclear since the teaching profession is listed separately. Despite these confusions, it is quite clear that the applicants were not overwhelmingly composed of 'intellectuals' and academics, but it does tend to confirm the tentative conclusion from the other data that 'white-collar' workers formed the greatest number of applicants.(33)

Another conception that the public have is that many so-called conscientious objectors were cowardly, selfish individuals, who were doing their best to escape an unpleasant task. It is evident from records of Tribunal decisions that the Tribunals did not agree with this point of view in a majority of cases. However, undoubtedly some men followed the procedures necessary to appear before a Tribunal, if only on the off-chance that the Tribunal would be taken in with whatever they had decided to use as the 'conscientious' grounds for their objections. A member of the public wrote to the 'Schoolmaster' in 1941 relating an episode that had occurred in his Local Tribunal:(34)

> The account of the teacher/conscientious objector who was complimented for bravery in an air raid which appeared in the 'Schoolmaster' recently reminds me by contrast of another type of objector who made strenuous opposition at his hearing before the Tribunal. As fast as they knocked down one fantastic objection he raised another, though he made no claim on the ground of conscience. At last the Chairman lost patience. 'Look here', he said, 'I'm not satisfied with your reasons for objecting. Now be frank. What is the real reason?' 'Well,' said the man, 'who wants to be a soldier while this .... war's on?'

But it is unlikely that the charming honesty of this applicant did anything to persuade the Tribunal to exempt him!

THE TRIBUNALS ASSESSED

Not surprisingly, the British public, and more particularly, Parliament, the press and organisations representing conscientious objectors, found much to say on the subject of the Tribunals. Two Local Tribunals evoked more criticism than any others: those of London and Newcastle. The CBCO reporter noted that, at the London Tribunal, the atmosphere was 'unfriendly', that the questions asked made applicants appear selfish or foolish, or that they were simply misleading, and the present or possible alternative employment of the applicant mattered more to the Tribunal than did his conscience. He commented angrily, 'This Tribunal is nothing but a labour exchange.'(35) He reported that the line of questioning would 'keep wandering away' from conscience to philosophy and politics. A PPU reporter thought that the questioning was so clever as to be almost intimidating at times, but admitted that it was invariably quite fair. The unfairness lay in the inability of the Tribunal to understand or sympathise with the pleas. He added that Judge Hargreaves, the Chairman, was occasionally unpleasant.(36)

From a less biased source came a more serious criticism of both the London and Newcastle Tribunals. In February 1940, F.W. Pethwick-Lawrence MP (Labour), a conscientious objector of the First World War, complained that (37)

> the proceedings in some of these tribunals are undignified,

> unseeming, and not really in accord with the wishes of the House of Commons ... instead of the judicial atmosphere which ought to prevail in a tribunal, there is a carping, bullying, brutal attitude taken up in them, which is not the one which commends itself to people who wish to see judicial decisions reached .... Three names were called out at the Newcastle Tribunal. The names of Donald, Cameron and Douglas were called out, and at once the Chairman Judge Richardson, remarked, 'Good fighting names. I think the holders of these names would turn in their graves if they could hear what some of these people are saying.' That is a very improper remark for a Chairman to make .... In the West London Court [London Local Tribunal], Sir Edmund Phipps called out in the midst of the proceedings, 'These miserable creatures'; and later, when they were speaking, he said, 'What tosh'. Surely these are not judicial remarks .... I do not think you are going to arrive at the truth as to their real mental and moral condition by shouting at them, by rushing a number of questions at them in a hurry, and by driving them into making foolish remarks which may or may not be their considered opinions.

Other complaints about the Newcastle Tribunal included that made by Pethwick-Lawrence that the Tribunal had ridiculed Jehovah's Witnesses on several occasions, and by the CBCO that Judge Richardson had insulted female conscientious objectors, although this was later, in 1942.(38)

Richardson was also severely criticised by several national newspapers. In April 1940 the 'Daily Herald' reported that Alfred Edwards, MP for East Middlesborough, was intending to ask for Richardson's dismissal as chairman of the Tribunal. The 'Herald' agreed with the MP. Richardson had, according to the paper, been making abusive remarks addressed to Jehovah's Witnesses. These included:

> I have the greatest contempt for your sort.
>
> You might pray and preach, but what good do you do?

The 'Herald' thought these remarks 'monstrous' and demanded that the 'responsible government department deal with him at once.'(39) 'Cassandra' of the 'Daily Mirror' was equally outraged at remarks made by Richardson later that year. Richardson had apparently said about Christadelphians: 'Middlesborough seems one of the centres of this poisonous body.' The columnist wrote: 'For sheer bad tempered, bigoted recklessness, allow me to introduce Judge Richardson .... This judicial swash-buckler regularly hits the news with his foolish and ill-timed jibes. His latest effort is well down to standard.' (40)

Some vigorous defence of the Newcastle Tribunal ensued in the House of Commons. David Adams MP (Labour), speaking in the same debate as Pethwick-Lawrence, said:(41)

> In so far as my experience in the North of England is concerned, regarding the tribunals to which reference has been made, I would say that they have been conducted fairly and dispassionately, and indeed, generously so far as judgements were concerned. I happen to know the judge, hailing from Newcastle, to whom reference has

> been made. I can only conclude that his irregular observations were merely a lapse, because he has a reputation of great fairness and consideration. We can only conclude that he was carried away.

This is not a good defence of Richardson's behaviour, since it was to ensure that an impartial atmosphere prevailed in the Tribunals, that county-court judges were chosen as their chairmen. It had been hoped that judicial figures with long court experience would not simply get 'carried away'.

In 1941 the third division of the Southern Appellate Tribunal (42) was much criticised by the CBCO, which, on one occasion, decided to ask the Minister of Labour to 'wind up' the division because of the 'deplorable record' of Sir Michael McDonnell, the chairman. The CBCO reckoned that his Tribunal was rejecting over 75 per cent of the appeals it heard, a figure they found remarkably high.(43) But it is dangerous to assume that, because an Appellate Tribunal was turning down large numbers of appeals, it was being unfair. It may just have been that the Local Tribunals feeding the Appellate Tribunals were getting their decisions right. The transcripts of the first division of the Southern Appellate Tribunal show it to have been courteous, giving the applicant every chance to speak his mind, and trying to understand his point of view, but nearly always deciding against him, not without reason. Yet the CBCO reporter at the Tribunal thought it 'appalling' and displaying 'no logic'.(44) Obviously the reporter had the advantage of being able to hear the tone of voice in which the questions were expressed, but even so, it is doubtful if he was really assessing the fairness of the Tribunal, rather than the number of dismissed appeals - two different considerations.

Tribunals were not only criticised for being intimidating, partisan or harsh. In a memorandum discussing the policy which ought to be followed on the question of appeals to Appellate Tribunals by the Minister, one civil servant in the Ministry of Labour and National Service commented that the South-Western Tribunal was 'notoriously lenient'.(45) This may have been a fair criticism, since Judge Wethered, chairman of the Tribunal, later admitted that the Tribunal had allowed too many unconditional exemptions.(46)

More general commentary on the work of the Tribunals came from various sources. An article in the Law Journal in 1940 stated that, (47)

> An inquiry into conscience is far too uncertain for a Court to undertake. It follows that the legal duty of a tribunal is to accept the applicant's apparently genuine conviction. To argue with him is outside its functions, and is both undignified and futile.

An article in the 'Manchester City News' complained that the Appellate Tribunals gave no reasons why they dismissed cases. It reminded the writer of the old sage advising a new judge never to give reasons for a decision because 'Your decisions will generally be right, your reasons will almost certainly be wrong.' The journalist felt that the reasons for turning down an appeal ought not just to be a matter of simple intuition.(48) The 'Daily Mirror' was critical of the cost of administering the Tribunals, telling its readers: 'So involved is the procedure ... that an expert authority has worked out that each man appearing before a tribunal costs the state £34. You pay. The Nation suffers.'(49)

However, by no means all commentary on the Tribunals was of a critical nature. One reporter for the CBCO repeatedly commented on the fairness of the South-Eastern Tribunal and of the Chairman there: 'The Conscientious Objectors have much to thank Judge Davis for at this Tribunal.'(50) The CBCO Board minutes record that the first sitting of the Northern Appellate Tribunal in December 1940 went well. It was operating courteously and with 'a sense of responsibility'.(51) A reporter for the 'Preston Guardian' commented that the Local Tribunal dealt with an influx of inarticulate farmworkers on one day with great patience. 'All the Tribunal's decisions ... reflected once again the toleration and fairmindedness of an English judicial body.'(52) In the House of Commons, Viscountess Astor told the House that she thought the Tribunals did 'extraordinarily good work'.(53)

> I know a woman, an American, who was married to a German, and who had been watching our tribunals before going back to America. She said to me, 'I had lost complete faith in human nature until I came back to England and watched the work of your tribunals. Since then I have regained my faith in mankind'.

In retrospect it is evident that the best judgment of the Tribunals and their work may be made by noting just how many of them escaped criticism from the many organisations and individuals who were keeping a vigilant watch on them. Much was made of a small number of patently unfair decisions in Tribunals, and a relatively small number of incidents or angry exchanges during the proceedings of the Tribunals. If a defence can be made for these lapses, it must be the extraordinarily difficult task for the objectors of convincing the Tribunals of their sincerity, and for the Tribunal members, of listening to and judging views with which they rarely had any sympathy or understanding. Obviously some applicants were bound to irritate the members. As Ernest Brown commented in the House of Commons: 'I would point out ... that some of those who give evidence are not themselves easy persons to deal with.'(54) An additional serious strain for both applicants and members was the constant threat of air-raid warnings. In one case, during an air-raid warning, the waiting applicants retired to an air-raid shelter nearby, while the Tribunal carried on its work, calling each applicant from the shelter when they were ready. It cannot have been easy either for the objectors or for the members to proceed, but proceed they did.(55)

If, then, in spite of all these difficulties, a large proportion of the Tribunals managed to carry on their work without serious criticism, it must be assumed that they were working adequately, if not well, under enormously trying circumstances.

Chapter 4

OBJECTORS IN CIVILIAN LIFE: the unconditionally and conditionally exempted objectors

INTRODUCTION

This chapter will concern those objectors who were exempted by the Tribunals from military service without conditions (a small number, as has been shown) and those exempted on condition that they undertook some civilian work under civilian control. By far the most applicants were awarded the latter or a 'B' decision at the Tribunals, a realisation of the fact that conscientious objectors were often more useful to the state in civilian employment than in a non-combatant unit in the Armed Forces where their contribution to the war effort could only be of limited value. No longer were wars fought by soldiers alone; the mobilisation of the whole civilian population had become a key factor in war management. If conscientious objectors could be found employment where a real contribution might be made, and which did not violate their consciences, the perfect answer to the problem of the conscientious objectors' role in a society at war had been found.

When an applicant was given the 'A' decision or unconditional exemption, the Tribunal's responsibility for him ended there. But for 'B' decisions, the Tribunals had to direct objectors to some sort of work. In theory that task seemed relatively easy: to find out from the government which areas of employment were short of manpower and to order that the conscientious objector find employment in that area. However, there were various snags to this arrangement. To begin with, the areas of employment most likely to require more manpower were often those which required specialised skills. Although both the 1939 Military Training Act and the 1939 National Service (Armed Forces) Act stated that the applicant given the 'B' decision should 'if directed by the Minister, undergo training provided or approved by the Minister to fit him for such work',(1) the Government was sceptical about training or retraining conscientious objectors while there were still many other people unemployed. Lengthy discussions on the feasibility of training conscientious objectors, especially for agricultural work, took place in the Ministry of Labour at the beginning of the war. At a meeting held in December 1939 the two representatives from the Ministry of Labour and the four representatives from the Ministry of Agriculture decided that, even if objectors were trained for agricultural work, they might not be able to find work afterwards. There was generally a surplus of agricultural workers and the Government might be criticised for wasting resources on a redundant project. 'It was generally agreed that the time for general consideration of training of conscientious objectors has not yet come.'(2)

In March 1940 the question was raised again as a result of a letter from Sir Miles E. Mitchell, a member of the North-Western Local Tribunal, to the Minister, Ernest Brown, asking that training schemes for conscientious objectors recommended for agricultural work should be set up.(3) The Ministry was well aware of the problem. In a memorandum a civil servant commented that the 2,000 objectors who had been told to obtain work in agriculture or forestry had found it almost impossible to obtain employment. Casual or seasonal work was easier to find but objectors were unlikely to be willing to give up their existing jobs for this, and anyway, during the periods of unemployment, they would be a drain on the nation's resources. Did the answer lie in training, asked the memorandum? Surely the ordinary unemployed ought to be given preference on any training scheme and in any case, objectors were unlikely to make agricultural labouring any sort of long-term career. Hence the money available would, in the long term, be wasted. Perhaps, the memorandum suggested, conscientious objectors could be trained at the various farming institutes grand-aided by the Ministry of Agriculture. The conclusion of the memorandum was that there was a possibility of setting up a training scheme for agricultural workers and including a number of objectors. But the matter was taken no further and large scale training schemes for conscientious objectors were never introduced.(4)

In addition to the fact that training would be needed for most of the industries which were short of manpower, those industries were often involved either directly or indirectly with the production of munitions. Tribunals could hardly recommend that objectors take up this kind of work. The other major problem for the Tribunals in recommending work for 'B'-decision objectors was that many employers were reluctant to introduce conscientious objectors on to their staff at all; Tribunal Chairmen and members were well aware of this and, later in the war, complained frequently of it. Unconditionally exempted objectors, of course, faced the same obstacle.

Members of Parliament had also addressed their minds to the problems of finding suitable work for conditionally exempted conscientious objectors. During the debate on the Military Training Bill the Minister was repeatedly pressed for an answer to the question of what exactly constituted 'work of national importance'. Ernest Brown, then Minister of Labour, said that the clause concerning this matter had been deliberately left vague because of the practical difficulties involved. But he added, 'the duty is placed on the Tribunal, and not upon the Minister, of settling what is work of national importance.'(5) This did not satisfy the MPs and Ernest Brown was eventually forced to make the assurance,(6)

> I hope that the Committee will be content to leave it to the Minister to see that the jobs are civilian and of national importance, and that he will either provide them himself or approve schemes for those who are sent on this work.

Parliament had in effect forced the Minister to take personal responsibility for the problem which, as Brown himself put it, 'bristles with difficulties'.(7)

Other points worrying MPs were cleared up by the Minister. He was asked if each objector would have to find his own job and answered, 'we feel that this is a burden which he will really have to bear himself.'(8) The Minister was also asked if a conscientious objector who was already performing 'work of national importance' would be required to leave that job for another. When the Minister replied that he would not, Duncan Sandys MP (National Conservative) wanted to make the point absolutely clear.(9)

> Did I understand ... that a tribunal would be able to fulfil paragraph (b) [the paragraph relating to the 'B' decision] by merely telling a man to go on at his own job on full pay?
>
> Ernest Brown: It might be in the national interest for that to be done.

It was clear that the Government had already taken a pragmatic view of the role of conditionally exempted conscientious objectors: why move a man from his own job where he was performing essential work? The tone of future governmental dealings with conscientious objectors had been set.

LAND WORK

Since by far the largest number of objectors awarded 'B' decision were directed towards agriculture, and since, in addition, some unconditionally exempted objectors had decided for various reasons to turn to the land for work,(10) it is important to investigate how many were able to find employment, how well they performed the work if they managed to find it, and what the views of other groups, the employers, the press and the public were of the newly recruited agricultural workers. There is much evidence to suggest that objectors had difficulty in obtaining work in agriculture at the beginning of the war. The 'Daily Dispatch' reported early in 1940 that objectors could not find the work prescribed for them. The paper pointed out that the onus was on the objector to find himself a job but that it was difficult because so many farmers were unwilling to employ objectors.(11) The 'Weekly Scotsman' also sympathised saying that forestry had received as many conscientious objectors as they could manage and that agriculture was in some places 'brimming'. The paper looked forward to the harvest when, it reasoned, farmers might need extra help.(12) At the beginning of 1941 the situation was still serious. The 'Evening Standard' reported that the Kent Labour Exchange repeatedly sent objectors to agricultural vacancies only for them to return saying that farmers would not employ them.(13) In June 1941 a conscientious objector wrote to 'Farmers' Weekly' complaining that he could not, hard as he might try, find a job in agriculture. He had applied for nearly forty situations without success and he knew others in the same predicament.(14) There were suggestions in Parliament in June 1940 that conscientious objectors were not making the effort to find work. Major General Sir Alfred

Knox MP (Conservative) asked Ralph Assheton MP, Under-Secretary at the Ministry of Labour, whether objectors looked for work and whether they got it. Assheton answered that they looked but could not always find work.(15)

> there are a number who have not been able to obtain agricultural work. County war agricultural committees are being asked to assist local officers of the Ministry of Labour as far as possible in finding suitable employment for them.
>
> Sir Alfred Knox: What is done in the case of a conscientious objector who is told to find work and cannot find it? Does he become a Cabinet Minister?
>
> Lieutenant-Colonel Ackland-Troyte M.P. (Conservative): Will my hon. Friend make it compulsory for all conscientious objectors to wear a white arm band, with the letters 'C.O.' in yellow?

Some pacifist projects sprang up at this time to provide work for conditionally exempted objectors ordered to take up land work and for unconditionally exempted objectors who, perhaps, had lost their jobs because of their views, or who wanted to work on the land for community service reasons. Henry Carter, a well-known Methodist pacifist(16) started the Christian Pacifist Forestry and Land Units in 1939 with the first unit, after consultation with the Forestry Commission, starting in Hemstead Forest in Kent. By the end of 1940, 400 religious conscientious objectors were employed in the afforestation units and they had spread north into Scotland and Wales.(17)

Even more community-inspired projects were started, ones not confined to objectors with religious convictions. In the autumn of 1940 a Community Land Training Association was formed to acquire a 300-acre farm in Lincolnshire where conscientious objectors with a vocation for agriculture were trained as community leaders. John Middleton Murry, by now editor of 'Peace News', saw the future of the country in terms of small communities working together in peace, and he was mainly responsible for raising an initial £10,000 for the land's purchase. Similar work was being done on the Adelphi farm in Suffolk where Murry was chairman. In the later years of the war, however, community projects for conscientious objectors reduced, partly because some of their keenest proponents had died (both Max Plowman and Eric Gill, two of Murry's closest friends and supporters, died at this time), and partly because most conscientious objectors eventually found work elsewhere.(18)

Despite initial difficulties, therefore, objectors were managing to find work on the land as early as the end of 1940. The Minister of Labour reported in Parliament in February 1941 that, 'up to 30th December last 86 per cent of the conscientious objectors registered for work on the land were reported as having obtained it.'(19) During 1941 the situation was further eased. The 'Western Morning News' reported that 'Devon's farmers' prejudices' were being overcome (20) and in June 1941 the 'Worcester News and Times' interviewed a Ministry of Agriculture spokesman who said that, 'On the whole, the scheme for putting conscientious objectors on the land is working out well.'(21)

There are two reasons for the increased availability of agricultural work for objectors in 1941. Firstly, the agricultural industry had, in some regions, developed a labour shortage in 1941. Farmers who in the past two years had either not been able or who had not been willing to employ objectors, were now forced to do so in order to maintain and increase levels of production. Secondly, outcries against objectors had lessened by this time, for many reasons. Some of them had proved themselves to be valuable workers in all types of employment. The press was beginning to write favourable reports of the contribution made to society by conscientious objectors. It was also much clearer after the Battle of Britain and the continual bombing of London and other areas of Britain that the civilian population ran just as much risk of death or injury as the Armed Forces. Objectors' well-publicised work in the Civil Defence Services often disproved the notion that all conscientious objectors were cowards.(22)

Another factor complicates the question of the difficulty or otherwise for objectors of obtaining work in agriculture at the beginning of the war. There was no general glut of workers in 1939; neither was there a general shortage from 1941 onwards. It depended very much upon the region in which the objector found himself. The Minister of Agriculture, R.S. Hudson, spoke about this uneven distribution of work in July 1940 when he was asked whether other counties, apart from Hampshire, had set up 'gangs' of conscientious objectors in order to do land reclamation work and other large-scale land operations. The Minister pointed out that the need for gang labour varied in different counties and that not many county war agricultural committees had as yet found it necessary to organise gangs on farms.(23) In those regions where work was available there was no 'story' for the press; it tended only to report incidents in which jobs were difficult to come by. Although the Ministry of Labour and National Service was talking about a 'black picture' in 1939, nevertheless, from their own inquiries which were then conducted at three-monthly intervals, it appeared that two-thirds of the objectors who had been told by the Tribunals to obtain work in agriculture had 'apparently managed to get it somehow'.(24) If two-thirds had somehow 'managed' in 1939, surely in 1941, when the situation had eased for the reasons mentioned above, very few conscientious objectors could have found great difficulty in obtaining work.

Not surprisingly there is conflicting evidence on how conscientious objectors performed land work, not surprisingly because they were neither more nor less than human beings showing the same human qualities of laziness, industriousness, courage, cowardice, greed and generosity as one would expect to find in any section of society. The objectors came from diverse backgrounds and possessed very different physiques. Labouring work came easily to some and not to others.(25) Work on the land was never meant to be a punishment to objectors for their refusal to join the Armed Forces. Rather it was meant to use the labour of these men in the most productive way. There is no doubt that in many cases these aims were not fulfilled; some objectors just could not settle down to their new-found occupation. There were complaints by the press and by the employers about the work of conscientious objectors on the land. The 'Manchester Daily Dispatch' asked in one article what could be done with 'deliberately slack' conscientious objectors. The article identified the 'slack' objectors as mostly elementary-school teachers and 'politicals' who, it had been hoped, would give up their stance when Russia entered the war, but had apparently failed to do so. The 'Dispatch' commented that it was 'no use' stopping their pay because they usually had a salary coming in from their 'whilom' employers anyway.(26) The 'Western Mail' reported in 1941 that, 'About a dozen conscientious objectors employed in a labour pioneer corps in South Wales doing agricultural work have joined the Army because they found the work too hard.'(27) In a letter to 'The Times' in 1941 which was reprinted in the 'Friend', a Major Carlos Clarke wrote that the conscientious objectors who were working on a drainage scheme in Surrey were 'inept'. Although they had no one to show them how to do the work, they were, in any case, unwilling to learn.(28) In 1944 an article in the 'Sunday Express' was entitled 'Farmers rage at sunbathing conchies.'(29)

There was indeed considerable resistance from farmers to employing conscientious objectors at all. Ernest Brown was aware of the problem in 1940. Speaking in the House of Commons he said to an MP who had just asked a question relating to the employment of conscientious objectors,(30)

> I would ask my hon. and gallant Friend to do one thing for me. If he will help me to break down the prejudice which farmers have to using conscientious objectors who are assigned to land work, he will help me greatly.

The Press Office of the Ministry of Labour noted a newspaper report of a meeting of the Dorset branch of the National Farmers' Union which passed a resolution stating that conscientious objectors should not be allowed to work on the land because they did not make good workers and that the farmers resented agriculture being regarded as an easy stop-gap to avoid military service.(31) 'Country Life' reported in the spring of 1940 that farmers did not want objectors working for them partly because they 'didn't know the job' and partly because the farmers viewed objectors' opinions with at least suspicion, not to say hostility.(32) Buckinghamshire farmers were also suspicious of objectors: 'these conscientious objectors have some kind of secret union.' They had the feeling that the objectors were trying to do as little work as they could.(33) In 1941 two papers reported an incident involving a farmer and a conscientious objector. The 'Star' and the 'Romford Recorder' noted that a farmer had been fined for punching an objector. The farmer maintained that the man would do no work.(34) In the same year the 'Daily Mirror' gave some considerable space to another meeting of the Buckinghamshire branch of the NFU at Aylesbury at which farmers complained that conscientious objectors 'wouldn't work' and were slovenly and rude. The farmers' feelings were well expressed by one member.(35)

> The Government treat us as fools. We can employ children and Conscientious Objectors on the land, and our own men are expected to endure a much lower standard of living than anybody employed in armaments factories, or practically anything else. Why? Because we don't stand up for ourselves. We are treated as fools all along. Why should we employ the second-class people? ... why should they [conscientious objectors] not be put sweeping roads or something. That is a good enough job for them. It is not what the Government are spending on agriculture, but what they are wasting on it.

The Bath branch had also felt it had reason to complain, passing a resolution in 1940, 'That we protest against conscientious objectors being set to work on the land as we consider it an insult to the farming industry.'(36) Farmers resented the fact that they had to pay objectors the same basic rate as other newly employed agricultural workers and so, according to the 'Wiltshire News', were happier when in the summer of 1940 the Agricultural Wages Board made an order modifying minimum wage rates payable to inexperienced workers. The rate was lowered from 48 shillings a week to 38 shillings.(37)

Despite these criticisms, however, praise was given from various quarters to conscientious objectors working on the land. The 'Preston Guardian', for instance, noted in 1941 that objectors were performing 'marvellous' drainage work, work which was very arduous. The paper felt 'bound to respect the principles of men who were prepared to go to these extremes to prove them.' These 'extremes' involved, among other things, wading waist deep into icy water.(38)

Objectors in Preston seemed to be pleasing the authorities generally. The Labour Officer of the County Institute of Agriculture felt strongly enough to write to 'The Times' about the way in which objectors had performed the work given to them by the Lancashire Agricultural Executive Committee. They had performed it 'well and willingly'. 'It is not easy to understand why difficulty has arisen in employing these men in other cases, but very much depends on the attitude adopted by foremen who are put in charge of the men.'(39) The 'Northampton Independent' reported Mr Dall, chairman of the Nene Catchment Board, saying that the objectors' work on the Nene drainage was by no means an easy task and that he paid high tribute to the way in which the objectors were helping to reclaim the land, and especially the Friends or Quakers.(40)

The Press was generally in favour of the idea that conscientious objectors should work on the land. The 'Spectator' thought it a good idea while hoping that it would not be spoiled by some people who were opposed to conscientious objectors.(41) The 'Worcester News and Times' published a letter which said that although farmers maintained that objectors were unfitted for land work, they could be usefully employed. 'If these men have satisfied a tribunal that they have a genuine objection and are granted some conditional exemption, it seems folly not to make use of their services on the land.'(42) And the 'Birmingham Evening Dispatch' reported the remarks of a member of the Midlands Tribunal who said that in some cases conscientious objectors were being exploited by farmers when they were paid no wages, just maintenance.(43) In 1941 the 'Weekly Review' summed up many people's view on what should be done with conscientious objectors. It considered the possibility that they could be put on to hard labour, but argued that this would be 'war-fever' at its worst. To leave them in their own jobs 'would be to create legitimate grounds for complaint by their fighting compatriots.' To work on the land, then, seemed a good alternative. They might not be very efficient in the job but 'at least they would learn something'.(44) As the Minister of Agriculture pointed out in 1944,(45)

> It is estimated that about 8,000 conscientious objectors are employed on the land, of whom the majority are in private employment. The output of conscientious objectors is necessarily dependent on such factors as their experience and physique, as well as their willingness to work.

OTHER AREAS OF WORK

Despite the fact that the Tribunals had not been given by the Ministry of Labour and National Service any real alternatives to agricultural and forestry as work of national importance for conscientious objectors, they felt able to recommend various other areas of work. Work in Civil Defence, either in the ARP (Air Raid Precautions) or in the NFS (National Fire Service) was recommended frequently. This again seemed a useful way of employing the manpower resources of objectors. Unfortunately, some of the local authorities which administered these organisations had made a decision not to employ conscientious objectors and this obviously hampered the chances of all objectors in civilian life finding or keeping work in that area.(46) Hospital work and ambulance work, also administered by local authorities and therefore fraught with the same difficulties, were also frequently recommended. Private ambulance services had differing views on the entry of objectors into their ranks. The 'Yorkshire Observer' pointed out in April 1940 that objectors were being refused permission to join the St John's Ambulance Brigade because of the views they held.(47) On the other hand, while the Friends' Ambulance Unit readily accepted conscientious objectors, it disliked the fact that they had been forced to take up this work as a condition of exemption.(48)

Later on in the war, coal-mining became a popular condition of exemption. The Board of the CBCO felt strongly that coal-mining was more closely related to the war effort than farming and it agreed at one of its meetings in 1943 that it should press the Minister of Labour to urge Tribunals never to give coal-mining as a sole condition of exemption.(49) By this stage of the war alternative conditions were quite common, for instance, 'Full-time work in agriculture or forestry under or approved by a public authority or full-time duties in Civil Defence or full-time work in a hospital as a stoker or porter.'(50) Soon after the meeting of the CBCO the London Appellate Tribunal amended a decision giving exemption on condition that the applicant undertook coal-mining work to exemption on condition that the applicant undertook coal-mining, land, hospital or ambulance work. The Board immediately decided to give publicity to this decision in its own publication, the 'Bulletin'.(51)

The Board continued to keep a watchful eye on the Tribunals to see if they were adhering to the decision made in the Appellate Tribunal. Hence, in March 1944, the Secretary of the CBCO wrote to Judge Finnemore of the Midlands Local Tribunal complaining that his Tribunal had been giving coal-mining as the sole condition of exemption from the Armed Forces. The government, wrote the Secretary, had made it clear that coal-mining was an integral part of the war effort and it should therefore never be given as the sole condition. The following month came a reply from Judge Finnemore. Beginning by saying that he could not conduct a public correspondence about Tribunal decisions and, therefore, asking the Board not to publish his letter, he explained his position on coal-mining for conscientious objectors in a number of points:(52)

1 I know what has been said about coal-mining. But of what activity has not the same been said? e.g. railways and especially farming without which all war work would stop? and income taxes and other taxes? We have been told more than once the whole war

effort would otherwise stop. Though I have not heard of one conscientious objector who has refused to pay his taxes.

2 I think coal-getting is only a border line case in the sense that every other industry is. If a man is a C.O. to getting coal, or growing food, is he entitled to use either? ...

3 In only one - or at most two - cases has a conscientious objection to coal-mining been put forward to us and we were quite satisfied that the real objection was merely an objection to an unpleasant job....

5 It is for the same reason that we have not usually prescribed any alternative, because our experience has been that in the vast majority of cases the easier alternative has been chosen. This has been another sad disillusionment for some of us.

In Parliament there were also disagreements about the issue of coal-mining. In 1943 Lieutenant-Colonel Sir Thomas Moore MP (Conservative) said that he thought conscientious objectors should be directed to coal-mining before anyone else. The Minister of Labour, Ernest Bevin, retorted,(53)

> The point is that a conscientious objector has a right to go to a tribunal, and coal-mining, like everything else, is one of the occupations to which it is open to the chairman to direct people. I cannot interfere with the process of the law.

A touch of realism was added to the whole coal-mining controversy when a Labour MP, Evelyn Walkden, commented in the same debate, 'in the mining villages we neither know nor care whether they are conscientious objectors or not. We want conscientious coal-getters to get coal, and that is all that matters to us.'(54) In any case, there is no further evidence that any more abuses of the Appellate Tribunal decision were made.

MPs had shown some discomfort at the thought that objectors, if they were already performing work of national importance, might be allowed to carry on in their own job and so fulfil their condition of exemption. One Tribunal at least discovered that it could remove some of the apparent unfairness of this decision by adding that an applicant must also carry on spare-time voluntary work as a condition of his exemption. A reporter in the South-Eastern Local Tribunal noted that one applicant was exempted on condition that he remained in his present occupation and that he continued his voluntary work with the Red Cross. It was the first time that this reporter had heard a case conditional upon spare-time voluntary work.(55) Presumably this particular applicant had no objection to carrying on voluntary work under compulsion. But the CBCO became concerned when, in 1941, it discovered that some Tribunals were making continued membership of the Home Guard an additional condition for exemption. The CBCO believed that this set a dangerous precedent for it was quite obvious that membership of the Home Guard was not civilian work under civilian control. A letter was sent to the Ministry asking for confirmation that 'the reference to the Home Guard in a condition is ultra vires and void as a condition, but only a recognition of the fact that the appellant is in the Home Guard, which the Tribunal approved.'(56) The Ministry approved of this statement and the CBCO decided not to pursue the correspondence but, nevertheless, to keep a vigilant watch on Tribunal decisions to make sure that membership of the Home Guard was never again given as a condition of exemption.(57)

It has been noted that those objectors, who were exempted on condition that they found work in agriculture, sometimes had difficulty in finding jobs, and when they found them, their performance in them was heavily criticised. The same general trends are found in other areas of work for unconditionally exempted objectors. Civilian prejudice against them was only to be expected and indeed many of them reported that soldiers were more sympathetic to their cause than the civilians they had to live and work with.(58) The Cooperative Society, for instance, apparently proud of its record of tolerance toward all manner of political and social views, in the summer of 1940 felt it necessary to dismiss objectors from its employment because 'they are detrimental to the trade of the Society'.(59) Similarly, the 'Manchester City News' published this piece in 1941:

> It is with regret that we announce that our principles and our policy have compelled us to part company with those of our staff who have declared themselves to be conscientious objectors.... There is no bitterness in this severance; we cherish the principles of others as we expect them to honour ours.

But the paper felt that it could not keep on editorial staff in wartime who were not 'whole-heartedly in the struggle'.(60) By 1940 an Anti-Conscientious Objector League had been formed in Blackpool and had begun the practice of sending out letters warning certain businesses that they had objectors on their staff.(61) Remarks made by Judge Richardson, chairman of the Northumberland and Durham Local Tribunal in Newcastle, while addressing the Hexham Rotary Club in 1942, that firms and businesses employing conscientious objectors should be 'cold-shouldered' were widely reported in the local and national press.(62) Lloyds Bank took a less drastic view of its employment of conscientious objectors. It announced that it would keep objectors registered in their present employment but, in order not to place them in a more favourable position than fellow employees in the Armed Services, their cost of living bonus would be deducted.(63) This was altogether a more satisfactory way of dealing with the vexing problem of employment of conscientious objectors.

Objectors wishing to go abroad also met with difficulties. In 1943 the Ministry of Labour opened a file on 'policy in considering applications from registered Conscientious Objectors for exit permits'. The question had arisen because a male objector had applied for an exit permit in order that he might take up a teaching position in Jerusalem working for the British Council. One civil servant commented that it was 'preposterous to allow a C.O. to go out to represent British culture abroad.' He felt that the Ministry would not want objectors spreading ideas not in harmony with the 'British ideal of total war'. The Friends Ambulance Unit were also interested in sending objectors abroad (64) and the servants of the Minister of Labour were not sure that the question of objectors working abroad was a matter for them at all. They were matters, they thought, of 'morale and prestige'. However, a decision was made to write to the CBCO stating that exit permits for conscientious objectors would each be considered on its own merits.(65)

In September 1940 the CBCO, increasingly concerned about employment for conscientious objectors, held a meeting with various other organisations to see what could be done. These organisations included the Friends Peace Service Committee, the Pacifist Service

Bureau, the Christian Pacifist Forestry and Land Units, the International Voluntary Service for Peace and the Community Service Committee.(66) There is no doubt that these organisations worked hard to relieve the employment difficulties of many objectors, sometimes providing the most suitable sort of occupation for those objectors who, while they felt unable to join the Armed Services, were most anxious to alleviate the distress that war brought upon thousands of people.(67) Benevolent societies such as the Hungerford Club, designed to help feed and clothe homeless and unemployed destitutes in London, welcomed the introduction of conscientious objectors on to their staff.(68) A few conscientious objectors were found employment assisting a research scientist of Sheffield University, Dr Kenneth Mellanby, in his experiments on the disease of scabies. The 'Sheffield Telegraph' commended the 'loyal cooperation' of objectors in Mellanby's work,(69) and the 'Daily Express' reported the comments of one objector who had been infected with the disease for experimental reasons. He thought that it was a task that conscientious objectors could do to 'help humanity'.(70) Two weeks after these reports had appeared in the press, Cecil Wilson MP (Labour) asked the Minister of Labour in the House of Commons whether there were any dangers involved for objectors in this work. Ernest Bevin replied that the occupation was entirely voluntary and that the objectors might suffer discomfort but not danger.(71) The research was apparently especially important because soldiers on active service were particularly prone to contracting the disease.(72)

THE FOLLOW-UP PROCEDURE FOR CONDITIONALLY EXEMPTED OBJECTORS

Of course it was pointless for the Tribunal to direct an objector into an area of employment as a condition of exemption if there were no means to ensure that he carried out the instructions of the Tribunal. Procedures for ensuring that objectors complied with their condition of exemption were included in the National Service Act, 1941.(73) Nevertheless, there was concern that some objectors were 'slipping through the net' and not making any serious effort to find the work prescribed for them. In February 1940 members of the London Local Tribunal wrote to the Ministry of Labour complaining that there was no 'effective follow-up procedure'. In its answer the Ministry agreed that the situation was 'worrying' but that work was being done to alleviate the situation.(74) In the same year there was concern at the Ministry that some objectors were using 'delaying tactics' to avoid finding work and some were described as 'wilful defaulters'. A meeting was held of all the District Commissioners of the Ministry to decide how to tighten up the procedures for discovering and dealing with this kind of objector.(75) By May 1941 a civil servant commented in a memorandum on this subject, 'the behaviour of some of the C.O.'s is becoming a scandal and we ought to try to put an end to it.'(76) The 'Evening Chronicle' complained in 1940 that there was little or no checking to see whether an objector had found work. (77) In the same year the 'Evening Standard' explained to its readers the procedure for making sure that objectors complied with their conditions but was doubtful if it was efficient.(78)

Another area of doubt lay in whether an unemployed conscientious objector was entitled to employment benefit. The 'Friend', the paper published by the Quakers, expressed its concern on this matter in 1940 when unemployment amongst objectors was particularly high.(79) People insured under the Unemployment Insurance Acts were entitled to benefit except in two special cases. The first was if a man had been dismissed from his job for reasons of misconduct or if he had left his job for no just cause. The second was when a man refused to take a job of which he had received notification by an Employment Exchange. In these cases unemployment benefit could be withheld for a maximum period of six weeks. The problems posed for the Ministry of Labour were these: if a man was dismissed from a position because of the views he held, was this 'misconduct'?; if he left his job because he conscientiously objected to some part of it, was this leaving his job 'without just cause'?; and if he refused to take a job notified to him by an Unemployment Exchange because he conscientiously objected to performing the work, was this a case where the Exchange would be justified in withdrawing his unemployment benefit? Naturally difficulties arose and a number of cases are described in detail in Denis Hayes's 'Challenge of Conscience'.(80)

It is difficult to assess just how many objectors in the Second World War were involved in cases like these. The only case which would affect conditionally registered conscientious objectors would be dismissal because of the employee's views. But it was clear by 1940 that the Umpire, whose decisions in these cases were final, had accepted the principle that it was not 'misconduct' to be a conscientious objector, even if the objector's views were so repellent to the employer that they led to his dismissal. Unemployment benefit could not therefore be withheld. The other cases are not applicable to conditionally exempted conscientious objectors because if they refused or left work to which they objected, they would automatically be sent back to the Tribunal for a reappraisal of the case, but unconditionally exempted objectors were affected.

In any case the problem of unemployment benefit was largely resolved by 1941 because of the drop in unemployment, and the National Service Act 1941 made the procedures for checking that conditionally exempted conscientious objectors were making efforts to find work considerably fairer and more efficient.

COMPULSORY CIVIL DEFENCE

On 26 March 1941 Ernest Bevin, the Minister of Labour and National Service, introduced the National Service Bill to the House of Commons. He said:(81)

> The object of this Bill is twofold. The first portion of the Bill introduces compulsory recruitment for the Civil Defence services on, broadly, the same lines as for the Armed Forces. The second part of the Bill proposes to make certain amendments of the National Service (Armed Forces) Act which experience has found to be necessary. The Bill has been found necessary owing to the shortage in certain districts of whole-time workers for Civil Defence .... This Bill accordingly proposes to make liable for Civil Defence service, men who are liable for service in the Armed Forces of the Crown. It imposes similar liabilities upon

> men who are registered under that Act as conscientious objectors, on condition that they take up some specified work of a civilian nature and will continue to be under civilian control, and it is particularly humanitarian. We have taken steps which will safeguard conscientious objectors against being drafted into the police because police have sometimes to carry arms.

Conditionally registered conscientious objectors, though not unconditionally exempted objectors, were now to be drafted into Civil Defence work. Various difficulties and objections arose from this new measure and they were fully aired in the House of Commons. John McGovern MP condemned the Bill: 'we think it is a thoroughly bad Measure in its intended effect upon conscientious objectors.' He stressed the case of the 'absolutists' who had been given exemption on condition that they undertook civilian work but who had been quite happy (82)

> to take land, forestry, land reclamation or drainage work, but they would refuse to go into Civil Defence work because they would then become part of the machinery of the war to which they had a decided objection .... Some people may be prepared to do work of a civilian character, but the fact that they are ordered to do so by a court and that penalties will be held over their heads will compel a large number of people to resist to the end.

One might have sympathies with the 'absolutists' who felt that Civil Defence was too closely allied to the war machine, but McGovern's second objection is illogical. The 'absolutists' who had only been conditionally exempt were already performing work compulsorily, and if they were not, then penalties were being 'held over their heads'.

In whatever way the 'absolutist' saw his predicament, if he was given exemption conditionally, he was far from free of state interference. As Bevin put it when the Bill was being considered in Committee, 'All that the Bill does is to make compulsory the obligation to one form of civil occupation.'(83) The real difficulty for most objectors not already performing Civil Defence duties voluntarily lay in the nature of the work. James Maxton MP explained their position well:(84)

> It is a big change to take a man who has been given exemption on condition that he does agricultural work, or that he works in a Friends' Ambulance Unit right into the middle of the recognised ARP services under orders and controls whereby he has lost command of his own acts.

Maxton could imagine the reaction of a Jehovah's Witness or a Plymouth Brother to compulsory Civil Defence duties: 'I am not taking my orders from Ernest Bevin; I am taking them from Almighty God, and from him only, and I will go through purgatory before I obey Ernest Bevin in place of God.'(85)

For those already performing Civil Defence duties voluntarily, the only objection could be that the work which had been done voluntarily should now become compulsory. It was the element of compulsion to which they objected, not to the nature of the work. Here again, though, they had already agreed to varying degrees of compulsion, for instance, they had obeyed the order of the Government to register and to appear before a Tribunal, and they had accepted that they would be compelled to perform some work, although there might have been an element of choice in the type of work.

The situation would have been far more difficult for objectors had not the Minister taken great pains in Parliament to reassure them and their representatives that there would be no abuses of the already agreed lines of administrative conduct. He promised that he would use his own common sense to administer the Act. Lewis Silkin MP (Labour) had commented in the Committee stage discussion that he wondered if the conscientious objectors were important at all in the Act. After all, they were already doing 'work of national importance'; why disturb them now? He went on to make another point:(86)

> many local authorities have refused to accept conscientious objectors for Civil Defence. But now these men will be forced upon them. I wonder how it will work, and how many of the men who are already doing full-time Civil Defence will work with conscientious objectors who are to be imposed upon them. Team work is very important indeed in Civil Defence.

With a show of impatience Bevin answered simply,(87)

> The discussion up to now has evaded the fundamental principle that I laid down .... Is Civil Defence a civilian job or is it a military job? ... In this Bill we take the greatest care to keep it a civilian job, and I cannot admit the principles of conscientious objection to civil work .... Subject to that, the rest is administrative ... I shall not move any man on essential work. For instance, agriculture is wholly reserved .... Allow me to use my common sense. Anyone who takes a conscientious objector away from essential work and creates trouble for himself somewhere else is not a very wise administrator. In connection with the Friends Ambulance Unit, nursing and hospital services, it is not my intention to disturb something that is working all right.

He did not, however, answer Silkin's second point on how the local authorities and their employees would receive objectors, and that, in many ways, was the key question. Indeed in January this problem had occurred to civil servants working on the Bill but they decided that if objections were made by local authorities, 'we will pay little attention'.(88) But by August the Ministry of Home Security had issued a circular to local authorities (89) stressing the Government's disapproval of those which refused to employ objectors purely because of the views they held. If there were objections from other employees to working alongside objectors, the position would have to be carefully explained to them. They should not deny the state 'in time of need the assistance of those who volunteer to serve it.' Most authorities accepted the spirit of the circular and there was very little difficulty after 1941 in the employment of conscientious objectors in Civil Defence.(90)

The newly called-up conscientious objector could express a preference for any of the Civil Defence Services in the same way that men going into the Armed Forces could express a preference for which Service they wanted to join. The choice was theoretically quite wide: he could elect to join the auxiliary fire service, the controls and reports service, the air-raid warden service, the first aid, casualty and ambulance service, the rescue service, the decontamination service, the messenger service or the police war reserve, although this last choice would be unlikely since it was well known that this service had, on occasion, to carry arms, and that the Minister of Labour had already promised that no conscientious objector would ever be required to serve in it. In practice the choice was not nearly as wide; it depended on which area the objector was in and how depleted the manpower of any one service was. In general it was the auxiliary fire service which needed most men and it was into that service that most were drafted.

It was obvious from the first that the Government and the Civil Service were making a great effort to avoid clashes between conscientious objectors and the state. For instance, rigorous instructions were given to the interviewing officers at the local offices of the Ministry of Labour. 'The object of the interview is to ascertain precisely on what work the men are engaged, how far it is essential, and whether they can be appropriately called up for full-time Civil Defence.'(91) Interviewing officers were told to see if the man was fit and, more importantly, to find out the man's general attitude toward Civil Defence work, whether he was actually anxious to perform it, whether he was not anxious but willing, and whether he was likely to refuse or not co-operate once he was in the Service. This was not to be done by direct questioning but by a general appraisal of the man's likely attitudes. The civil servants working on the Bill predicted that the Government would be asked why unconditionally registered conscientious objectors were not to be made liable for call-up into Civil Defence. The answer was to be that it would be a mistake to force Civil Defence on someone who objected to having anything to do with the war effort. But they stressed objections by conscientious objectors with conditional exemption were not allowed. Allowing objections to work just because the state ordered it 'would result in political and social chaos and the disappearance of an ordered state.'(92) A circular printed in 1943 reminded local authorities and local offices of the Ministry of Labour that the Government had pledged itself not to make conscientious objectors handle munitions work or anything to do with the war effort:(93)

> The Minister's wish is, broadly, that persons who profess conscientious objection should make as full a contribution as possible to the national effort, but that they should not be required to carry out duties which are closely connected with, or allied to, the military side of the war effort.

Despite the Government's conciliatory and practical attitude, there were 3,156 prosecutions of conscientious objectors for refusing the medical examination for Civil Defence up to the end of 1948. Of those, 2,829 were convicted. However, fines were not prohibitive and prison sentences were rare.(94) There was an uneasy stalemate between the Government's determination to make conditionally exempted objectors contribute to the Civil Defence of this country and the stubborn conviction of many conscientious objectors that compulsion was not the way to do it.

Chapter 5

PUBLIC EMPLOYERS AND THEIR ATTITUDE TO THE EMPLOYMENT OF CONSCIENTIOUS OBJECTORS

INTRODUCTION

When in 1940 the local authority for West Penwith debated the question of whether or not it should retain the services of conscientious objectors, an astute Councillor commented:(1)

> About 95% of the population would like about half of these conscientious objectors to be put on a ship one dark night, landed on the north coast of Europe and left to their own resources, but the difficulty would be to know which half to pick out.

Local authorities discussed the question in some depth, partly as a result of the strength of public opinion, and partly because Councillors themselves often felt very deeply about the matter. But there appear to have been very few debates in which everyone was agreed that there should have been no 'conscience clause' in the National Service Act of 1939. Rather, Councillors were proud that Britain should be so generous towards its recalcitrants. It was this very sort of freedom for which Britain was fighting. However, while the principle of a 'conscience clause' was generally accepted, there was a lurking, and very natural, suspicion that many of the so-called 'conchies' were shirkers, cowards and 'funks'. And it was this dilemma which dominated many of the local-authority debates on the taxing question of employment of objectors in the early stages of the war.

One of the reasons for the occurrence of the debates was quite simply that local authorities were paying out public money to their employees, and this was money of a public which, especially by 1940, was whole-heartedly engaged in total war. The wave of unpopularity which conscientious objectors suffered in the spring of 1940 naturally included revulsion at the idea of paying rates which would go to subsidise the easy-living 'conchies'. Refusal to pay rates until conscientious objectors were dismissed was not unknown,(2) and in the Northwich district, rather than pay rent to the collector who was recently registered as a conscientious objector, several women walked miles to the Rural Council offices to pay.(3) In an area of Manchester feeling was so strong that Council offices where a conscientious objector worked were stoned.(4)

But other, even more pressing, events forced the Council debates.

Threats of strike action from local-government services such as the ARP and the Fire Brigade because members of those services felt they could not work alongside conscientious objectors put even the most liberal-minded Councils in a virtually impossible situation. It was blackmail at its most effective, since the cessation of any of these services would be disastrous for a community in wartime. Apparently in one village, firemen were so enraged because there was one conscientious objector in every four fighting men that they were moved to write to the parish council registering their horror and suggesting that it was all due to the influence of the local parish clergy. (5) Councils simply could not afford to upset firemen by the retention of the services of conscientious objectors. Another crucial dilemma was whether or not to retain the services of conscientious objector teachers, and public feeling about this question was at its most incensed. A separate section will be devoted to that issue.

These pressures, in addition to opinion in the Councils, promoted the debates, which were invariably well attended, noisy and long. The questions with which the Council were faced were threefold. Was the dismissal of conscientious objectors in public service, firstly suitable, secondly practical, and thirdly legal? Was it suitable to dismiss men who had very often been working efficiently and loyally for some years in the Council's employ, but who now held opinions which, although contrary to the spirit of public opinion, were acknowledged in law by Parliament? Was it practical to dismiss men from services, many of which were short of manpower, and to engage men who did not have the experience of the original employee? And was it in fact legal to dismiss contracted employees, who might belong to the National and Local Government Officers Union or to the National Union of Public Employees, and whose interests had in any case been protected in an Act of Parliament? In the emotionally charged atmosphere of many of the debates, some of these questions were never identified or answered; decisions were sometimes arbitrary and unreasoned, and only a detailed examination of the debates will show how this delicate and fraught subject was handled by local authorities.

THE DEBATES

There is no doubt that many local authorities had debates on the question of dismissing conscientious objectors. In fact some authorities debated the subject hotly for no apparent reason. For instance, it was reported in the local newspaper that the Cheadle and Gatley Urban District Council 'last night declined by eight votes to five to rescind its resolution that conscientious objectors should be dismissed from the council's employment although informed that none of its employees was a conscientious objector.'(6)

In order to debate any subject in Council, a resolution had to be moved by one of the Councillors. Very few resolutions were worded exactly the same, although there was one, the Lytham St Annes' resolution, which became particularly well known and was often presented to other Councils in its original form. It called for, among other things, the retention of conscientious objectors on condition that they accepted soldiers' pay only.(7) While most of the resolutions were, from the conscientious objector's point of view, damaging to a greater or lesser extent, there were some resolutions calling for the retention of conscientious objectors on full pay and with war bonuses. At a meeting of the Essex County Council, for instance, in November 1940, Councillor Captain Romanes justified his resolution to retain the services of all their conscientious objector employees by saying that as the Government had acknowledged them, so should local authorities. The local paper reported,(8)

> Captain Romanes said in his resolution was laid down a future policy which the Council would follow .... If they discussed a man, or refused to employ him, because he held views which were not popular with those who had the power of dismissal and appointment, they were going exactly the same way that Germany went in the case of racial troubles. If a man had a conscience he could no more go against his conscience than a Jew could help being a Jew.

The Council finally decided that it would only dismiss conscientious objectors if their work was adversely affected by their beliefs.

There were, then, at least five possibilities for resolutions to be moved on this subject. They could demand complete dismissal of objectors; dismissal for the duration of the war, that is, 'leave without pay'; continued employment but on soldiers' pay and without promotion; continued employment without war bonuses; and continued employment on full pay and with war bonuses, and promotion opportunities not affected.

The discussions on the motions are interesting and enlightening because they give a fair indication of the many responses, some emotional but some reasoned, that there were to conscientious objectors in Great Britain during the Second World War. The experience of the Great War had conditioned many to accept the existence and, indeed, inevitability of there being a 'conscience clause' in the National Service Act, and to recognise that certain numbers of people would take advantage of it, but this did not of itself relieve a concern that there were men in responsible civic employment who would refuse to serve their King and Country in such an emergency. Further, even genuinely sympathetic Councillors felt that it was unfair that objectors might gain financial or promotional advantage over their colleagues in the Armed Forces, though they sometimes wondered if the authorities which they served had the right to interfere with a question that had already been decided in Parliament.

At their worst, the debates degenerated to an almost farcical floor show for venting vicious and poisonous comments on the character of conscientious objectors.

> It is time we took some steps to get rid of these leprous sores. (9)

> I have no time for C.O.'s. They are like worms and should be tramped upon.(10)

> Send them to the pits!(11)

> They should be shot at dawn.(12)

> They have a yellow streak.(13)

All this publicity is making conscientious objectors into plaster saints with putty haloes.(14)

There was also a curious debate at Lanark about the introduction of horse flesh as a suitable wartime meat for public consumption. Major Basil Monteith 'twitted members on their dislike of horse flesh', and won 'first prize for fatuity' in 'Reynolds News' when he remarked to the County Council, in an effort to show them the worthiness and honour in eating horse flesh: 'I think it is a horrible idea that a noble animal like the horse should be eaten by conscientious objectors.'(15)

But that sort of comment was usually inspired by loss of patience in a heated debate, and there were far more serious and thoughtful attacks on objectors and their role in society. An often repeated contention was that objectors should be dealt with in the same way as those in Germany or at least that they were lucky not to be. One Councillor commented: 'Men who register as conscientious objectors should be treated as they are in Germany. They should be put in concentration camps.'(16) The idea that Hitler 'would know how to deal with C.O.'s' (17) implied admiration for the decisive way in which the Chancellor dealt with reluctant participators in the German war effort. The apparent unfairness of the objectors' position upset other councillors, the principle being that 'the men who risked nothing should not be allowed to profit at the expense of the men who did so.'(18) A Gloucester Alderman asked:(19)

Is it fair or right or just for us to encourage young men to run into funk holes when other brave men are going in their tens of thousands to protect us, our country, the Empire, our mothers, wives and children?

While many shared these views, and expressed them, their speeches were often concluded by remarks like, 'but we can't have Hitlerism here.'(20) Their worry was more the relationship of a local authority decision to that of Parliament. The chairman of the Liberal Committee in Herefordshire asked: 'Are we, a little tin-pot County Council, going to over-ride what the Government has decided?' (21) The opposite view was expressed by a Councillor in Esher: 'The Government may have recognised objectors, but it doesn't follow that local authorities have to.'(22) The legality of dismissal will be discussed later; suffice to say that there was some question in councillors' minds about whether their authorities had the power to dismiss objectors, and even if they had, whether there was a moral defiance in not acting exactly in accordance with central Government's and Parliament's wishes. The autonomy of local authorities in this matter was sometimes in doubt.

The suitability of dismissing objectors from Council employ therefore largely depended on the personal opinion of each councillor, for, as members of the public themselves, they might feel horror either at having objectors working in public employment or at dismissing men purely because their views did not coincide with those of the Government. One councillor in Welwyn Garden City said that (23)

The law has recognised human consciences and for the Council to take action in another direction would be a retrograde step unjust to those people and would not do anything more than express a

> feeling of sentiment which was in the minds of a certain number of people.

Conversely, another councillor in Wimbledon was worried about what other countries might think of Great Britain were her local authorities to retain objectors in their employ: 'We are asked what America will think of us. They will think it is the biggest bunkum in the world to allow objectors to do as they are doing.'(24) Most Councils decided that it was suitable to dismiss objectors for moral reasons, although there were Councils who thought that it might be dangerous to allow objectors to work for them in that they might interfere with other employees and therefore constitute a threat to the safety of the state. 'Defeatist' talk was thought by the Government to be, in 1940, a real danger, but, judging by the fact that Government Ministries continued to employ objectors, it did not see them as part of this threat. Local authorities might well have used this argument to bolster their already existing prejudice against employing objectors because it was logically possible that they could have spread dissatisfaction and discontent especially if they were political objectors. It is difficult to ascertain just how evangelical most conscientious objectors were.

The impracticality of dismissing objectors was raised in an article in the 'Manchester Guardian' in August 1940. The writer stressed that the objectors could be of service to the community, and then wryly driving the point home, commented:(25)

> Could a proposal be adopted, perhaps, to dismiss the conscientious objectors after the war when we may need their work less urgently? By then, of course, we shall have all cooled off, and nobody will want to persecute them.

There was, for instance, a serious shortage of teachers, but local authorities persisted in dismissing those holding undesirable opinions. The interesting point is that despite the lack of manpower, authorities felt that it was so suitable, or morally correct, to dismiss objectors, that they took the risk of being undermanned and understaffed. Alton Rural District Council had a well-publicised debate on whether or not to employ a badly needed additional Sanitary Inspector who was also a conscientious objector. The Council, in fact, turned down a request from the Ministry of Health to reconsider their original decision not to employ the man. The letter from the Ministry pointed out that there was a shortage of trained Sanitary Inspectors and that every person had been given a legal right to entertain a conscientious objection to military service. In the face of this central Government request some Councillors appeared to waver, but the chairman argued that the applicant was unsuitable on grounds other than his conscientious objection. The following exchange ensued:(26)

> Clerk: I wonder why, he is a qualified Sanitary Inspector?
>
> Chairman: That doesn't cover everything .... Do any of us think that a conscientious objector, who has to do public work in this district, would be able to do it properly? Would anyone work with him, or would he be accepted in the district?
>
> Major Wessel: He will find himself in an ash-bin, sir.

Mr Joy: Ought we not to look at it from the point of view of national manpower?

Major Wessel: From the nation's point of view he ought to do his job as a soldier. (Hear, hear.)

The local paper commented:(27)

Whatever one's views on the matter may be, it does seem that from the point of view of the national effort, a conscientious objector who is doing a useful job of work is going to be of more assistance in running the war than one who, because of his views, is ostracised by the remainder of the community.

Alton was not convinced, however.

In rare instances councillors declared themselves to be conscientious objectors, and then Council had to decide whether or not they could allow one of their own number to continue working for the authority. A Labour objector councillor from the Haydon Park Ward was most indignant at his forced resignation:(28)

The Council might just as well have said that those members who opposed the purchase of Wimbledon Park Golf Course should be removed from the council because they might have to serve on a committee charged with administering the golf course.

The public had voted this man into power, and he was therefore, until the outbreak of war, presumably considered to be a hard-working, responsible and useful member of the community. But the abhorrence at his personal opinions far outweighed the value of his service to the local authority. There was a very large moral gap between disapproval of the purchase of a golf course and disapproval of the war.

The legality of dismissing objectors from employment was perhaps the most widely debated issue in the papers, the Tribunals and the public at large. A journalist for the 'Widnes Weekly News' thought that it was possible that Widnes Town Council had left itself wide open for a successful legal action: 'The law of the land definitely defines the position of the Conscientious Objector. I have yet to discover a Town Council has power to nullify parliamentary legislation.'(29) In a letter to the editor, the secretary of the St Albans Division of the Liberal Association commented in the Barnet Press: 'We all know that Barnet is a very important place, but it will be news to all of us that the Urban District Council wields an authority superior to that of Parliament.'(30) But these were value judgments rather than accusations of illegality. The Esher Councillor was, of course, wrong when he said that, while the Government had recognised objectors, it did not mean that local authorities had to. The law of the land recognised objectors and their right of exemption to military service, and so local authorities were bound to recognise them too, as they would be bound to recognise any law. The crucial issue was whether there was any legal sanction against dismissing men from employment because of their conscientious convictions. However undesirable the Government may have thought the action of local authorities, it at no time claimed that their action was illegal.

Government disapproval emanated from the highest quarters. Churchill himself had stated that: 'anything in the nature of persecution, victimisation and man-hunting is odious to the British people.'(31) But this speech was made in 1941 when many of the decisions had already been taken, and had therefore not influenced the early debates. It was also made in a slightly different context from that of the position assumed by local authorities. He was announcing in Parliament that the BBC was lifting its ban on the right of objectors to broadcast. However, it might well have played some part in causing certain authorities to vary the severity of their decisions, for instance, from complete dismissal to dismissal just for the duration of the war. Ernest Bevin's comments were more direct. In August 1940, he remarked in Parliament: 'I am not being assisted in this difficult and technical job by the action of a number of local authorities, who seem to take the law in their own hands.'(32) He was replying to questions regarding Fenner Brockway's plan for objectors to give to a central fund the money they earned above that of a soldier's pay. Also in August Alexander Sloan MP (Conservative) asked the Minister what action he intended to take concerning the dismissal of a man from his employment due to his pacifist views. Bevin replied: 'I have stated previously that I strongly deprecate action of the kind alleged, but I have no authority to control an employer's action in such cases.'(33)

In May 1941 Ernest Bevin was informed by George Strauss MP (Labour) that conscientious objectors in local authority employment were still being dismissed and 'persecuted'.(34) A few days later Strauss brought the matter up again. Bevin announced that he had nothing further to say on the issue except that 'no doubt this further question by my hon. Friend and my Reply will assist in giving publicity to the matter.'(35) Government Ministers obviously disapproved of the actions of local authorities. Herbert Morrison, Minister of Home Security, was an ex-conscientious objector himself. But Ministers were powerless to do more than disapprove. In late 1941 the Minister of Health, Ernest Brown, was asked by an MP whether he proposed to intervene on behalf of the Medical Officer for Dudley, who had been suspended for the duration of the war because he had registered as a conscientious objector. Brown stated in a written reply that: 'The appointment, dismissal or suspension of the Medical Officer of Health of a County Borough does not require my approval, and I have no authority to intervene between the Council and their officer in this matter.'(36) A Government circular was issued to all local authorities concerning dismissal of objectors and it will be discussed in detail later. It should be said here only that, although it was a strongly worded document, it was merely a request and not an order to lift any ban on objectors working in Civil Defence.

Chairmen on Local and Appellate Tribunals repeatedly commented on the actions of local authorities. It was a common occurrence for Tribunals to exempt men from military service on condition that they continued in their present employment, and for those men to return to the Tribunals reporting that they were not able to fulfil this condition because they had been dismissed. Judge Frankland at the Manchester Local Tribunal said: 'It is unfortunate that even such an august body as the Cheshire County Council take upon themselves to interfere with the law in this way and attempt to disturb the whole balance' (37) and 'I do wish these public bodies would not interfere with the law.'(38) The chairman of the North Wales Local Tribunal referred to the Swansea local authority when he heard that they had dismissed an objector: 'Most people in authority in Swansea seem to have lost their heads altogether.'(39) Sir Miles Mitchell, chairman of the Northern Appellate Tribunal, commented in May 1941 that a conscientious objector was recognised by law, and that it was 'stupid' of local authorities to decide that they would not employ objectors, but more significantly went on to make it clear that he realised that this action was not illegal however frustrating and 'foolish' it might have been.(40)

While there was no law to prevent any employer dismissing employees whose opinions he did not agree with, Regulation no. 2 in the National Service (Armed Forces) (Prevention of Evasion) Regulations, 1939 (No. 1099) provided that:(41)

> An employer shall not terminate the employment of any person employed by him by reason of any duties or liabilities which that person is or may become liable to perform or discharge by virtue of the provisions of the Act.

A prosecution was brought before a Huddersfield stipendiary magistrate by three conscientious objectors, who had been dismissed from their employment, against their former employer. It was suggested by the prosecution that the employees were dismissed by reason of duties or liabilities which the employees were or might become liable to discharge by reason of the provisions of the Act. The liabilities and duties to which the Act referred were the process of registering and submitting to medical examination, and appearing for duty in the Armed Forces when required to do so. The magistrate at Huddersfield found that the dismissals were solely on account of the conscientious objections of the employees, but in any case as an article in the 'Solicitor' points out: 'there is no obligation under the Act upon any person to register as a conscientious objector, this being merely a right. The obligation is simply to register.'(42) There was some discussion among lawyers about when a man officially became a conscientious objector. A letter to the editor of the 'Manchester Guardian' stated that:(43)

> In the eyes of the law until a man holds a certificate of registration as a Conscientious Objector from a tribunal he is not a Conscientious Objector but a potential soldier. As such he is protected by these regulations from dismissal and after he is called up his reinstatement in due course becomes obligatory.

But if, as was the case, it was legal to dismiss men on account of their views, then the legal quibbles become irrelevant. The only stumbling block which remained was the terms under which an employee took up his employment, that is, whether the contract of employment had been broken. Some objectors fought their dismissal on these grounds but were generally unsuccessful. Local authorities' contracts of employment would obviously be drawn up by their own professional solicitors and as long as the required amount of notice of termination of employment was given according to the contract, the authorities, when taken to court, had no case to answer.(44)

Close study of local authority debates revealed no discussion of the actual legality of the decisions that were made in them. Local authorities were well aware that they were on safe ground there. What worried councillors was whether they were abiding by the spirit of the law which had given the right of conscientious objection; but there was no law which protected objectors' rights of employment. The issue was again moral, but not legal.

The value of the debates in determining the course of action for local authorities to take is in itself debatable. They certainly represented a wide spectrum of opinion and belief, and were a reliable barometer of public feeling on the issue of conscientious objectors. Some were conducted reasonably, coolly and thoughtfully, and some became the scenes of rowdiness and uncontrolled vindictive behaviour. As the 'Hampshire Herald' pointed out: 'The subject of C.O.'s is one that can usually be relied upon to provoke a fairly heated discussion whenever it is raised.'(45)

THE DECISIONS

Once a resolution had been debated, a vote was taken and the decision made. It has been explained that there were various forms of resolutions moved, and the majority of those voting in favour or against the motion would depend on the content of the resolution and the unity of the Council. The West Riding County Council made headlines in July 1941 when it managed only by fifty-one votes to forty-nine to reaffirm its decision to suspend for the duration of the war all objectors in its employ. On the first occasion the margin had only been one vote. The local paper said that: 'Events once again proved the Council to be almost equally divided on the subject.'(46) When votes resulted in large majorities, the decision was usually one of dismissal not retention. A paper asked: 'Should C.O.'s receive the same privileges as men serving in the Forces? That question was answered by Leyton Council on Tuesday with an emphatic and expressive 'No'.'(47) The voting had been almost unanimous. Since conscientious objection to military service was still not by any means a socially acceptable stand to assume, there were bound to be more councillors who felt strongly against objectors than for them. Where decisions were made in favour of objectors it was not usually through any sympathy with the ideal, but through respect for the lead given by central Government or the feeling that to dismiss men for their views was tantamount to Hitlerism.

The most famous resolution was that of the Corporation of Lytham St Annes, Lancashire, which was passed in April 1940. It was: 'that in the opinion of this council Conscientious Objectors should be compelled to carry out work of national importance on rates of pay no higher than, and under conditions no better than, those of H.M. Forces.'(48) The Corporation was anxious for this resolution to be adopted nation-wide, and went to great pains to publicise it and to invite other authorities to follow suit. When the resolution was presented to the Chadderton authority, the Chairman looked at it and commented: 'It'll do no harm' (49) and the Council passed it without discussion. It seemed to be a very suitable compromise. In fact, as became clear when Ernest Bevin considered a compulsory scheme for withdrawing pay earned by objectors over and above the pay of soldiers, the proposition was hopelessly impractical. While soldiers' actual pay was low, the perquisites that went with the job were innumerable. Apart from free board and lodging, free clothes in the form of uniform, free transport, free laundry, free or half-price entertainment, all contributed to make life in the army, at least in financial terms, less difficult. The Lytham resolution mentioned conditions of work which were more crucial. Sitting in an office and fighting in the trenches were obviously very different occupations, but even this point must be qualified. Offices in certain parts of the country ran a considerable risk of being bombed, and Government services, such as the Auxiliary Fire Service, and Ambulance services were dangerous occupations in wartime, whereas not all soldiers were on the front line; some postings were, without doubt, much safer than civilian life in Great Britain. The real unfairness lay in the fact that an objector, in his accustomed civilian employment, would be able to gain the experience which led to increased pay and promotion. The soldier, on return to his old job, would be in exactly the same position as when he left it, except that he would be so many years older. Yet the same applied to those who were in Reserved Occupations, and those who were unfit, medically or otherwise, for active service. It was never suggested that these people should be penalised in the same way. It had to be assumed that they would have wanted to join the Armed Forces had circumstances allowed them, whereas objectors obviously did not, and would therefore have to pay the price for that unwillingness.

This situation did not of course apply to all objectors. Many were soldiers in the Non-Combatant Corps, others were uprooted from their pre-war jobs to work on the land or in forestry or in Government services such as the ARP or AFS. In this way they could not benefit at the expense of the fighting forces. The problem only affected those who were exempted on condition that they continued in their present employment either because it was a Reserved Occupation or because it was considered to be work of national importance, and many objectors who worked for local authorities were exactly in this position. Conversely, there were a number of objectors working for local authorities because Tribunals had exempted them on condition that they should do so, and later when the Government took powers to direct them to that work. They were therefore losing the same promotion prospects in their own jobs as men serving in the Armed Forces. The whole question was a taxing and complicated problem, and local authorities who had no wish to dismiss objectors, still had to consider the fairest action they could take. If they deducted pay from all objectors in their employ, some would be paying twice for their conscience, and yet not to make some attempt to reduce the financial and promotional position of objectors who were permanent employees compared with that of a soldier seemed patently unfair and ungrateful. As the Bolton municipal worker wrote angrily to his local paper when he learned that when he was called up, he would suffer financially, whereas a worker who was an objector would be allowed the full war bonus: 'One is led to the astonishing conclusion tht the local authority is prepared to grant better treatment to conscientious objectors than to men willing to fight for their country.'

Perhaps the fairest decision that a local authority could take was to bar a conscientious objector from promotion until the end of the war, but few authorities felt that they could limit their action to that extent. Temporary loss of promotion prospects were always coupled with financial penalties, be it withholding war bonuses, or reducing pay to that of a soldier. Some authorities withstood the pressures of public opinion completely and retained and employed objectors on full pay, with promotion prospects and war bonuses. The number that took this course was, however, small.

TABLE 1 The decisions of the local authorities

	County and City Boroughs	County Councils
No decision or not reported	24	25
All objectors dismissed from service	31	19
All objectors dismissed for duration of war	14	6
All objectors retained on soldiers' pay only	8	8
All objectors retained on full pay with or without war bonuses	18	5
TOTAL	95	63

Table 1 shows the decisions of the local authorities. The interesting point is that in both the County Council and the City and Borough Councils, the percentage number of Councils which dismissed their objectors either permanently or temporarily is almost equal. Altogether 66 per cent of County Councils, and 63 per cent of City and Borough Councils dismissed their objectors.(51) But the large numbers of Councils which cannot be accounted for are crucial, since Councils which did not debate the question either retained their objectors or did not have any in their employment at all. If they retained them, it would radically alter the validity of the preceding figures. Considering the remarkable strength of public opinion in 1940 and 1941 concerning the entire issue of conscientious objectors' place in society, it is unlikely that the local press would fail to report a debate on the subject in the district's Council. For the same reason it is unlikely that Councils with objectors in their employment would not debate whether or not to retain them. It may be tentatively assumed, therefore, that Councils for which there are no press reports of any debates, simply did not have any objectors in their employment about whom to make a decision. Over half the decisions that were made, consequently, cost objectors their jobs, and as a decision to dismiss objectors was invariably coupled with a decision not to employ them, unemployed objectors (often unemployed because their original employers had taken the same decision as local authorities) would have to look further afield than their local Councils for employment.

TEACHERS IN THE EMPLOY OF LOCAL AUTHORITIES

The decision whether to dismiss another large sector of public employees, the teachers, was usually taken separately by the Education Committee, but was generally in line with the policy of the

Council on all its objectors. Public outrage at public servants who were conscientious objectors was largely directed at the teachers. A letter to the editor of the 'Hemel Hempstead Gazette' represented a widely held view:(52)

> As a serving soldier I object to my children going to a school at Hemel Hempstead at which there are conscientious objectors on the staff, as these so-called men might teach my children, and they are not fit to do so.

The 'Daily Sketch' reported in April 1940 that:(53)

> Parents in the colliery village of Bersham, near Wrexham, are protesting against a conscientious objector being allowed to teach at the village school. Many parents are threatening to take their children away from the school unless the teacher is withdrawn.

The teacher-parent relationship is, at best, always a delicate one, and parents are often asked to swallow huge doses of teaching methods and disciplinary measures of which they are fearful and distrustful; but the idea of allowing their children to be taught by men and women with views so alien to the spirit of the war effort, was, for many parents, a medicine not just unpleasant, but unpalatable.

Feeling was also strong in the House of Commons. In April 1940 Lieutenant-Colonel Acland Troyte MP (Conservative) was certain that if conscientious objectors were allowed to continue teaching there would be 'a danger to the country of the spread of views held by these people.' The President of the Board of Education, Herwald Ramsbotham, replied:(54)

> I have no reason to suppose that the teachers concerned would fail to observe the principle, to which the teaching profession itself attaches great importance, that political propaganda should in no circumstances be introduced into the schools.

In August 1940, W.J. Anstruther-Grey MP (Conservative) asked the President whether parents were still being 'compelled' to have their children taught by conscientious objectors (55) and later that year, Sir James Mellor MP (Conservative) asked the President to limit the number of objector teachers. Ramsbotham was unmoved:(56)

> The Board has never sought ... information as to the private convictions of teachers in a matter of conscience; nor am I prepared ... to call into question the discretion which rests with local education authorities and governing bodies of schools in making appointments to their teaching staffs.

Still some MPs were dissatisfied. In July 1941 Sir James Mellor commented: 'if conscientious objectors are genuine they are likely to be of a fanatical disposition and if they are not genuine they are even more undesirable for the purpose [of teaching].' Major General Sir Alfred Knox asked the President: 'Does the right hon. Gentleman intend to keep conscientious objectors misleading the young?'(57)

Responsibility for children's education was shared by local authorities and many were in agreement with the parents and with the Conservative MPs. Fife County Council rejected an application from an objector for a teaching post on the grounds that he should not be allowed to 'distribute the doctrines of pacifism among the youth of Fife.'(58) The difficulty was that there were many other unpopular views held by teachers for which school boards and local authorities did not feel they had to penalise teachers. If all teachers had had their moral and political views examined because the authorities were worried that they might indoctrinate their pupils with them, the number of dismissals would have been ridiculously damaging to education in Great Britain. Lord Mamhead, a distinguished public servant, ruefully pointed out in a manifesto on the future of religious education in schools in 1941:(59)

> while in many parts of the country teachers who are Conscientious Objectors have been turned out of the schools, because they have lost the confidence of parents, declared agnostics are allowed to remain. No-one seems to mind if an atheist is teaching children.

The point was that the zeal and enthusiasm for the winning of the war was so great that pacifist views were by far the most unpopular one could hold. With Russia fighting with the Allies, even Communism was not subject to the bitterness and ridicule which it received before and after the war.

So while the decision to dismiss teachers was made on quite understandable emotional grounds, attempts were made to rationalise the action. Councils would claim that they were dismissing objectors on 'educational grounds', the grounds being that the teacher, or in the case of the Lancashire Education Committee, the headmaster, would occasion so much resentment from parents, pupils and other staff, that they would be forced to resign anyway. Had the local authority supported the headmaster, this might not have been the case, but Lancashire was determined to dismiss.(60) In a debate at North Riding, an alderman asked whether a conscientious objector would be considered more fit and proper to deal with children after the war. Presumably if indoctrination was going to occur, it would occur after the war as well as during it. A councillor replied it was hoped that the objector would be a sadder and wiser man when the war was over. (61)

The Board of Education was in the same position as the Ministry of Labour and the Home Office. Teaching had been made a Reserved Occupation after the age of 30, but even so there was scarcity of teachers, so while the Board could take no legal sanctions against the actions of local authorities, it made it quite clear that it did not approve. In July 1940 it was moved to issue a circular to all local authorities drawing attention to the circular which had been issued by the Ministry of Home Security and the Home Office.(62) The contents of this and other Government circulars will be dealt with in detail later, but the gist of it was that it was undesirable to dismiss any employees for reasons of their personal opinions. It carried no legal status however. An MP asked the President of the Board, Herwald Ramsbotham, in Parliament whether, now that the Government position had been made clear, the Board would reinstate teachers who had been unfairly dismissed. He answered: 'As the House knows, the authorities have the power to engage, employ or dismiss teachers. All I can do is to issue principles for guidance, and that I have done.'(63) In May 1940 the Ministry of Health had also issued a circular which, it was generally agreed, was easily misunderstood.(64) Apparently it had led to some authorities conducting 'heresy hunts' as in the 'notorious' case where teachers were requested to sign a declaration:(65)

> I hereby solemnly and sincerely declare that I am not a conscientious objector, or a member of the PPU. Nor do I hold views that are in conflict with the purpose to which the Nation's effort is

directed in the present war, and I further declare that I wholeheartedly support the vigorous prosecution of the war.

When a man who had been asked to sign this document presented it at the North Wales Local Tribunal, the Chairman commented: 'You would be rendering a public service if you would forward this document to the Ministry of Labour. I question the right of a public authority to request employees to sign such documents.'(66) Publicity in the press and in the House of Commons soon put a halt to this practice, but it showed again that there was nothing the Government or the Tribunals could do about the actions of local authorities except to request them not to dismiss teachers and act in an advisory capacity.

The largest teachers' union, the National Union of Teachers, sympathised with the dismissed teachers, but was none the less powerless to protect its members. In August 1940 it passed a resolution welcoming the Home Office circular which it hoped would clarify the position and would help safeguard public servants against unfair or capricious dismissal by local authorities.(67) Newspapers repeatedly pointed out the fatuity of dismissing teachers, stressing the fact that authorities even dismissed teachers who were over 30 and were therefore not liable for call-up. In fact, this was a logical step. Authorities were basing their right to dismiss on the grounds that the pacifist views might be taught to the children, and men and women over 30 were just as likely to do this as those under that age. One journalist asked to whom the right to decide the ratepayers' interests belonged. A conscientious objector might also be a very good teacher, and by dismissing him, authorities would be damaging the pupils' education.(68) The truth was that, for some authorities, there was no price too high to secure the dismissal of conscientious objectors from teaching in schools.

THE GOVERNMENT LEAD AND CIVIL DEFENCE

At the beginning of the war the Government gave no lead to local authorities on the question of the employment of objectors; it might not even have envisaged that there would be a problem with public employers, although Neville Chamberlain seemed to anticipate some difficulty when he stated in May 1939, as he was introducing the Military Training Bill: 'I want to make it clear here that in the view of the Government, where scruples are conscientiously held, we desire that they should be respected and that there should be no persecution of those who hold them.'(69) Nevertheless, it became a serious issue as the war progressed into 1940 and 1941. In various speeches Ministers made it clear that, while they disapproved of the authorities' actions, they had no power to stop them. The first clear lead came in the Circular 1522 issued by the Home Office and the Ministry of Home Security in July 1940. It was 'for the guidance of Local Authorities in dealing with complaints made against employees in local government service reflecting on their fitness to remain in the public service.' Although the words 'conscientious objector' were not used in the memorandum it was evident that this was one type of employee to whom it referred: 'The first principle to be observed is that in this country no person should be penalised for the mere holding of an opinion, however unpopular that opinion may be to the majority.'(70)

This did not appear to influence many local authorities in their decisions however. It is difficult to find specific mentions of it in any debate. However, in August 1941, a new circular was issued by the Ministry of Home Security concerning the 'Enrolment of Conscientious Objectors for Civil Defence'.(71) This was a very clear statement of the policy which the Ministry wished local authorities to pursue. So frank was the statement that it must have been somewhat of an embarrassment to certain local authorities which had taken a strong line against objectors. It said that despite Government recognition of conscientious objectors, 'authorities in some cases have nevertheless hesitated to accept [them] as members of their Civil Defence Services notwithstanding that in many places such persons have been accepted and proved satisfactory in every way.' It went on to say that the National Service Act of 1941 has altered the position 'materially' and 'it would be in accordance with the policy of the Government that authorities should not in future refuse to accept, as volunteers for civil defence, persons who hitherto have been regarded by them as unsuitable candidates.' Stressing the efficiency, and even bravery, of objector Civil Defence workers, the Minister concluded that while authorities would still have discretion over whom they employed, 'in exercising this discretion they should not, in the Minister's view, be influenced solely by the consideration that the person concerned is a conscientious objector.'

Reaction from local authorities was swift. Croydon Town Council, for instance, which had dismissed all its objectors, despite grumbles from some councillors that objectors were now joining the Civil Defence units to escape military service, now decided to allow objectors into their Civil Defence services.(72) There are many other cases of change of policy although it is impossible to say how many exactly; by the end of 1941, conscientious objection was not the great issue it had been. Papers now preferred to comment on the bravery of objectors rather than their cowardice. However, it is likely that most local authorities changed their general policy towards employing objectors in Civil Defence because there were, in addition to the circular, two other factors to influence them. Firstly, the shortage of manpower in the Civil Defence services was acute, so the practical argument for employment of objectors was reintroduced with a vengeance. Secondly, the behaviour of the objectors already in employment had convinced not only the authorities but, more importantly, the other Civil Defence workers, that not all objectors were 'shirkers' and 'funks'. ARP workers who had refused to work with objectors now had a very different attitude. The Chairman of the Birmingham ARP Committee commented: 'Conscientious Objectors have carried out rescue work of the most dangerous kind without trepidation.'(73) The 'Hornsey Journal' noted the London County Council decision to lift their ban on employing objectors in Civil Defence services. The ban had originally been imposed because members of the service said they could not work with conscientious objectors:(74)

> Then came the raids, in coping with which conscientious objectors had to take a full share. The result has been a big diminution in the number of those who were reluctant to work with them. This has made it possible for more to be employed.

Not every authority was persuaded however. When presented with the

Government circular, Enfield Fire Brigade reported to the Council that they had no recommendation to make on the matter,(75) and the 'Barnet Press' noted that despite the circular the Finchley Borough Council still refused to employ a Quaker conscientious objector as a full-time ARP warden.(76) And it is only fair to add that while many authorities now allowed objectors to work in Civil Defence, many felt no obligation to lift their ban on their working in other civic capacities.

THE ACTIONS OF OTHER PUBLIC BODIES

Local authorities were by far the largest of the public employers, but the Civil Service and such bodies as the British Broadcasting Corporation employed a considerable number of people, and therefore warrant some attention. In August 1940 there were 272 objectors employed by the Civil Service.(77) In July of that year a resolution was passed by the Executive Committee of the Association of Ex-Service Civil Servants asking the Government to dismiss objectors whom they described as a 'potential menace to the state'. The President said: 'They should be turned out. They are sowers of discord, discontent and disloyalty.'(78) Three days later it was announced in the House of Commons that objectors in the Civil Service would be debarred from consideration for promotion. MPs had asked that these men should not secure any advantage over those in the Forces. The only rider to this was if a Civil Servant refused to accept the decision of the Tribunal, his service with his Government department would be terminated at once.(79) So, in effect, objectors were retained in the Civil Service with no disadvantage except that they would lose their promotion prospects until the war ended, in order to keep in line with their colleagues in the fighting forces. There was little press comment about this measure except in the 'Sunday Dispatch' which in an article complaining about the inefficiency of the Post Office commented: 'You may be interested to know that the Post Office employs more conscientious objectors than any other department in the Civil Service.'(80)

In December 1940 the BBC decided to sack all objectors working for it and stopped a broadcast of the Orpheus Choir because its conductor, Sir Hugh Robertson, was a conscientious objector.(81) 'If and when they obtain non-combatant service with the Forces the Corporation will be prepared to consider applications by them for reinstatement.' (82) But in March of the following year Churchill announced in Parliament that the BBC had lifted the ban.

He said that a conscientious objector would be able to broadcast in his capacity, for instance, as a musician: 'But I think we should have to retain a certain amount of power in the selection of music. Very spirited renderings of Deutschland Uber Alles would hardly be permissible.'(83) But none of the public authorities were given nearly so much publicity in their treatment of objectors as the local authorities however and it is the public's attitude to the decisions that will conclude this chapter.

PUBLIC OPINION AND ITS INFLUENCING FACTORS

Pressure of public opinion was one of the main reasons why local authorities had to debate the question of whether or not to retain the objectors in their employ. It is always difficult to discover by what process the public forms an opinion and how far the press, the Government, the various churches or the trade unions affect it. The press presumably would argue that it represents and expresses public opinion, and to some extent tries to influence it. Certainly the local press was the vehicle for the feelings and opinions of its readers through its letter columns. One dismissal which aroused a particularly bitter controversy was that of Dr Peate, Keeper of the Department of Folk Culture and Industries at the National Museum of Wales. He had been given, at his Tribunal, unconditional exemption and had promptly been dismissed by the Council for the Museum. Immediate public outrage produced the statement from the Council that Dr Peate had not been dismissed on the grounds that he had registered as a conscientious objector. A letter to the editor of the 'Western Mail' asked why had Dr Peate been dismissed if it was not for his conscientious objections?(84)

> Is it the authorities who have lost their heads and suddenly gone mad? Or can it be that Dr. Peate has suddenly lost his grip and made a mess of his department? Has he lost his character or lost his health? It surely can't be that he has broken some priceless pottery of the Stone Age by some clumsy and crude experiment in excavation. Surely he can't have dug up a modern bungaloid and mistaken it for an ancient Welsh cottage.

The Council merely announced that Dr Peate had 'seriously misled it on a crucial issue' and that an earlier statement by him in which he said that 'he was prepared to do his duty to his country' 'has now proved to be an untruth.' Another contributor to the argument asked: 'Does anyone who is concerned with decency and justice in this country and with the counter-acting of Nazi ideas, doubt that a full enquiry into the administration of the so-called 'National' Museum is urgently necessary?'(85) But there were a number of contributors who supported the Council in their decision. The members of the Council, it was said, were entitled to the 'freedom to say that they don't wish to maintain Dr. Peate and such like people in public posts for which they are paid public money.' Another wrote on similar lines and signed himself as 'Freedom For All'.(86)

Paying out public money was a recurring theme in the campaign to dismiss objectors from local government employment. In one debate a councillor remarked that he was sent to the Council by 700 ratepayers and not one of them would agree to the Council 'paying away' their money to conscientious objectors.(87) At Coventry, when the ratepayers heard that the Council was going to retain its objectors, a 'very noisy' protest meeting was held (88) and residents at Crookesmoor would not pay their rates until objectors were taken off the Corporation.(89) Papers varied greatly in their approach to the subject but quite a common ploy was for papers to cover themselves against accusations of being unpatriotic, as in the Evening Standard- 'Let us make ourselves clear. We oppose the pacifist case in this war.'- and then go on to complain about the action of the local authorities: '[but] it is morally indefensible for another body in effect to override the judgement of the tribunal.'(90) Another journalist from the same paper protested at the right of Councils to employ objectors on soldiers' pay: 'I do raise the question whether local authorities are justified in fishing like this for service on the cheap.'(91) Other papers balked at having to publish letters supporting objectors at all, but commented that it was this very freedom for which other men were fighting. If any generalisations can be made, the national press was probably less biased on the issue than the local press.

The trade unions varied in their approach as well. While the NUPE was moved to put forward a resolution at the Trades Union Congress in 1940 strongly deprecating the actions of certain local authorities in dismissing objectors,(92) some branches of NALGO, representing the 'white-collar' worker, pressed their Councils not to employ objectors. The Norwich Branch was very upset when it was suggested that there were thirty objectors who were members of NALGO in Norwich. A vigorous denial of this accusation was published in the local papers. (93) The Church of England gave no strong lead either. The Archbishop of York said that the dismissals were 'utterly deplorable and in the deepest sense unpatriotic' (94) but local clergy were sometimes at the forefront of the battle to rid Councils of the presences of objectors. A socialist and an objector had been elected Chairman of the District Council of Potters Bar; the vicar there, who led the opposition to the appointment, wrote to the 'Freethinker' saying:(95)

> It might quite well be that Councillor Osgathorp would make an excellent Chairman, but that our civic head in this critical year [1940] should hold the opinions which he openly professes, is entirely out of place and offensive to a great majority of citizens.

The Church as an institution supported the war, but there were bound to be differences of opinion within it on the subject of conscientious objection.

Probably no one influence on public opinion was greater than the influence of events themselves. The year 1940-1 was dominated by the Fall of France, Dunkirk, the Battle of Britain and devastating bomber attacks. As Mass Observation, an opinion poll, rightly pointed out, public opinion hardened, had to harden, in every way.(96) The stolid determination to survive, and in the end, to win the war at all costs could not be anything but detrimental to popular opinion of conscientious objectors. The local authority dismissals must be seen as a natural and understandable reaction to the enormous wartime pressures which rested on every individual. Many Councils remained amazingly objective about the issue, some even went so far as to sack employees who would not work with objectors, a councillor commenting optimistically that: 'The duty of men employed by the Council is the ratepayers. Personal opinions of this kind should not enter into their employment.'(97) But personal opinion was bound to enter into the question of employment of conscientious objectors, and it was only the passage of time and events which lessened the strength of feeling on this once explosive issue. A journalist anticipated the loss of interest: 'this question of the employment of conscientious objectors in public services is now an old controversy. It has been discussed ad nauseam in Council Chambers all over the country.'(98)

By the end of 1941 the public was settling down to a long, grim but calmer period in the war. While conscientious objectors and the organisations representing them continued to fight their own battles for what they considered to be fair treatment, the public and local authorities representing them devoted their energies to other topics.

Chapter 6

CONSCIENTIOUS OBJECTORS IN THE ARMED FORCES AND IN PRISON

THE NON-COMBATANTS

If a Tribunal was satisfied that an applicant objected only to combatant service, it ordered that he should undertake non-combatant duties in the Armed Forces, usually in the Army. At the beginning of the war the procedure was that, once a man had been given this 'C' decision, his name was removed from the provisional register of conscientious objectors and placed on the military-service register with the proviso that he should be called up for non-combatant duties only.(1) The normal procedure for entry into the Armed Forces then followed; the man was medically examined and in due course called up for service.

In 1939 there was only one truly non-combatant corps, the Royal Army Chaplains' Corps which was obviously not suitable for anyone except ministers of religion. However, there were some corps which were accepted as performing work of a non-combatant nature. Corps in this category included the Royal Army Medical Corps, the Royal Army Dental Corps, the Royal Army Pay Corps and the Royal Army Veterinary Corps. The first of these was attractive to many conscientious objectors. It offered the opportunity of humanitarian work and the Tribunals, despite the fact that there was absolutely no compulsion upon them to order anything but that an applicant should undertake non-combatant duties in the Armed Forces, often added a recommendation that the applicant should be placed in the RAMC. The motive for this action varied. In some Tribunals it sprung from a genuine desire to enable conscientious objectors to perform the most suitable sort of work, in others it was used as a bait to attract those objectors who had applied to be exempted completely from service in the Armed Forces but who were anxious to carry out work of a humanitarian nature. Many objectors did not realise that the Tribunal's recommendation was only a recommendation, not an order. During the course of 1940 the RAMC was finding it difficult to accept any more recruits; the corps was beginning to become overmanned. It was then that many conscientious objectors were disappointed and found themselves in other corps with less obvious humanitarian commitments. At this time some Tribunals warned objectors that they could only recommend their applicants to the RAMC and that the actual possibilities of gaining entry to it were slim. Others continued to recomment it without comment either through ignorance of the state of the RAMC or through the misguided belief that it was their duty to recruit as many conscientious objectors into the Armed Forces as they could.

As early as November 1939, the Secretary of State for War, Leslie Hore-Belisha, had warned Edmund Harvey, Independent Progressive MP for the Combined Universities, that there was no guarantee that objectors recommended for the RAMC would be able to enter it: 'That depends on whether there are the necessary vacancies for them.'(2) Exactly a year later Anthony Eden, one of Hore-Belisha's successors at the War Office, answered a question in Parliament from William Glenvil Hall MP (Labour). Hall had asked if non-combatants who wished to transfer to the RAMC would be expected to undertake combatant duties. Eden replied that it would not be necessary for a man to renounce a conscientious objection in order to be transferred to the RAMC but warned, 'I should add that the RAMC has already a large number of conscientious objectors.'(3) The implication was clear. Possibilities of transfer within the Army to the RAMC, let alone direct entry from civilian life by conscientious objectors were slight. By the time that the Non-Combatant Corps was created in April 1940 the RAMC was gaining a reputation as a reception centre for conscientious objectors. A member of the Corps wrote to his home-town newspaper that all RAMC members were being dubbed 'conchies' (4) and a volunteer in the Corps complained in a letter to the 'Daily Dispatch' that,(5)

> It is galling to think that places which should be filled with men who have given their spare time in peace time to practise first aid work are taken by mealy-mouthed Pacifists who have done no voluntary work in this direction.

Even if objectors were successful in entering the RAMC the problem remained for all non-combatants: what duties would they be expected to perform? Until April 1940 none of the corps they were entering was officially non-combatant and so it was especially necessary to know which duties within any corps were permissible and suitable for non-combatants. At the start of the war there was some confusion about this matter. For instance, when civil servants discussed the problem in 1939 they thought that those in the RAMC would have to learn to use a rifle as a split, and to unload a wounded man's rifle.(6) Yet this was certainly handling 'lethal weapons', a term that had crept into usage as some sort of guideline as to what was combatant and what was not. The Secretary of State for War did not give a definite answer on this point when in January 1940 he replied to a letter from Cecil Wilson MP (Labour) asking him to define non-combatant duties and to say which corps were non-combatatant. The Minister confirmed that there was no non-combatant corps apart from the Royal Army Chaplains' Corps. He observed (7)

> that the meaning of 'non-combatant' depends a good deal on the point of view of the person using it.... In short, their [conscientious objectors'] attitude depends on the general character of the work they will have to perform and not on any official ruling as to whether the work is combatant or non-combatant.

The truth was that there was no official ruling at all on the precise meaning of the term 'non-combatant', perhaps rightly so. With all the 'multifarious duties', as the Minister put it, that there were in the Armed Forces, it would be an impossible task to define which were combatant and which were not. However, there was room for compromise. Some guidelines were issued by the Army Council at the time of the formation of the Non-Combatant Corps for the training and employment duties of non-combatants in the Corps. These were as follows: (8)

Training

All personnel of the Non-Combatant Corps will be given training in:
(a) Foot drill, without arms.
(b) Physical training.
(c) Passive air defence.
(d) Anti-gas measures.
(e) Decontamination of rearward areas.
(f) Specialist duties as required including cooking and clerical work.

Employment

The personnel of the Non-Combatant Corps will be employed, in addition to normal administrative duties, only on general duties appropriate to their category, including:
(a) Construction and maintenance of hospitals, barracks, camps, railways, roads and recreative grounds.
(b) Care of burial grounds.
(c) Employment at baths and laundries.
(d) Passive air defence.
(e) Quarrying, timber-cutting, filling in of trenches.
(f) General duties, not involving the handling of military material of an aggressive nature.

While there were still particular problems after this, the main worries about duties seem to have been allayed.

Aims for the administration of the new Corps were set out in an official note issued by the War Office in April 1940. It first praised the courage and bravery of many non-combatants in the last war and said that it was essential that the Corps 'should build a high standard of morale and esprit de corps.'(9)

> The Non-Combatant Corps will be provided with officers and N.C.O.'s from the A.M.P.C. [Auxiliary Military Pioneer Corps] who will be attached for duty with the Non-Combatant Corps. There will necessarily be a difference of personal view on the ethics of combatant service for the country between the permanent and the attached personnel of the Corps; but this should not prevent their willing and active co-operation in making each Company a first-class unit of which all can be proud.... Officers and N.C.O.'s whatever their personal views, will ensure that there will be no discrimination or victimisation of any kind.

The connection between the NCC and the AMPC soon gave rise to some confusion and not a little resentment. Sir George Broadbridge MP (Conservative) asked a muddled question in Parliament about conscientious objectors in the Pioneer Corps. As the Financial

Secretary to the War Office, Richard Law, pointed out, Broadbridge confused the AMPC with the NCC, but it was obvious that he was trying to say that members of the AMPC, largely composed of ex-servicemen, resented their connection with the NCC.(10) In May 1941 'The Times' printed a statement issued by the War Office: 'Doubt appears to exist in the minds of the public as to the relative status of the Pioneer Corps and the Non-Combatant Corps (Conscientious Objectors).' The statement explained that the Pioneer Corps was formed mainly of veteral soldiers and reservists who had volunteered for further service.

> On the formation of the NCC - a conscientious objectors' corps - it was decided that as the only use that could be made of these men, who are unarmed, should be for labour, they would be attached to the Pioneer Corps for administrative purposes only, while specially selected officers and N.C.O.'s from the Pioneer Corps should be put into command of these companies.

It explained that the Pioneer Corps was a combatant unit and its members could be called upon to bear arms, while the members of the NCC could not. The NCC was an entirely separate unit, its members did not wear Pioneer Corps badges, 'and, although used for many of the jobs allotted to the Pioneers, are attached to that corps only for their general organisation and training.'(11)

All kinds of opinion were expressed about the NCC, both by observers and members. Some criticised; an article in one newspaper said that the NCC was arousing resentment. Its members capitalised on half-price entry into places of entertainment, YMCA and Services canteens. They wore uniform very similar to those of the fighting forces and the Home Guard. The Home Guard, it was felt, would want to 'prevent such men from masquerading as soldiers and, above all, misappropriating soldiers' privileges while remaining unwilling and unprepared to share a soldier's responsibilities and risks.'(12)

War diaries kept by senior officers of the companies of the NCC show that 'continuous difficulties [were] experienced by company officers and N.C.O.'s through verbal expressions of opinion regarding Conscientious Objectors made by combatant unit officers and Ordinary Recruits as well as [the] general public in [Great] Yarmouth.' 'Considerable tact' was needed in handling the situation. (13) And later, 'General praise from Departments employing N.C.C. labour with regard to willingness and energy for work but much opposition to principle of conscientious objection. Discipline of Company very good.'(14) But in the village of Gamlingay in Bedfordshire the commanding officer noted, 'Villagers are antagonistic to N.C.C. and openly say so.'(15)

But others also found much to praise in the work of the NCC and non-combatants generally. In 1960 a newspaper review of a book written by Major-General Sir Brian Taylor about the blitz quoted a few lines about the work of conscientious objectors in bomb disposal: 'Particularly worth praising were the Conscientious Objectors-about 160 of them - who volunteered for this dangerous work. They did it jolly well, too.'(16) The 'Empire News' reported admiringly that 250 young Scots soldier conscientious objectors had transferred from the NCC to the Royal Engineers so that they could undertake bomb disposal work.(17) Bomb disposal work demanded a certain courage which excited admiration from everyone. But it was illogical to single out bravery by conscientious objectors in bomb disposal work from other equally courageous bomb disposal soldiers as it was to single out the misbehaviour of some conscientious objectors in one company. It would have been as erroneous to judge that, because some soldiers performed courageous acts in the war, all soldiers were courageous, or to deduce that, because some soldiers deserted from the army during the war, all soldiers were cowards. However, the difference was that conscientious objectors, because of the stand they were making, were, in many ways on trial and exceptional behaviour by any of them, either bad or good, influenced their contemporaries' view of them. It would have taken a supreme act of objectivity then, as it does now, not to judge a minority group by the activities of a small number of them.

More interesting,then, perhaps, is the view the non-combatants took of their work, its usefulness and its possible frustrations. In 1942 an article was written about the NCC and was submitted to the 'New Statesman and Nation' for publication. The paper refused to publish the article on the grounds that it was too long and 'not convincing enough'. From the tone of the article it would appear that the author either had been or still was a member of the NCC and therefore an obvious bias must be taken into account.(18)

> There is a widespread ignorance about this rather anomalous institution, an ignorance which was fully shared by most of its members until they were posted into it - and even after that.... An Army Council Instruction sets out, rather vaguely, the kind of work on which the NCC should be exclusively employed, and in practice it falls to Company Officers to decide what can be done. One NCC Company was engaged in clearing up work in Coventry, one section was set to clear up in a factory producing aeroplane parts; they refused and all were put on charge. But the section was soon put to another job, and the charges were dismissed. In other respects the authorities have not been so scrupulous. In about half the Companies pressure has been put upon men to transfer to combatant units. This has taken two forms; petty persecution and brow-beating. The persecution may sound very petty; restricted leaves and passes - few men had more than one 48 hour per year, until recently, when matters have improved, and sleeping out passes have been extremely rare; endless kit-inspections in spare-time; unnecessarily unpleasant conditions of work and living. Soldiers will know how oppressive these factors can be in the conditions of Army life.... There was a Captain who came on parade with his revolver toyed with it lovingly, and announced that he'd like to shoot the whole bloody lot of the men standing before him at attention.... In a few Companies there has been no pressure, rather the reverse. In others methods used have been ones of bribery rather than bullying. In one Company any man who would transfer to the Pioneer Corps was promised a stripe immediately. It must be noted that no promotion is possible in the NCC.... In spite of this general willingness to work hard, which has continually impressed the authorities, the men have not hesitated to go on strike when they have felt keenly on some point of condition.... There have been cases where NAAFI girls and WSV people have refused to serve NCC men in canteens. On the whole civilians have been more hostile than people in the Forces, and one Company had to be moved because the civilians with whom they were working protested....

> In all cases the tendency has been for the atmosphere to change from one of hostility when the NCC arrived, to one of appreciation before they left.... Where there's NCC men there's books. At every break during working hours half the men are reading. The standard of education and intelligence is very high, far higher than in the average run of Army units. Many of the men have been to Universities and the vast majority have taken pains to educate themselves. In civilian life most of the men are either students and teachers or professional and clerical workers; there are comparatively few artisans or manual workers. The educational standard of the men often puzzles the officers and NCO's; it worries them to see education and pacifism going hand in hand.... The politicals want to obstruct at every turn; the Plymouth Brethren ... will not oppose any authority at all.

Much of the article's contents can be confirmed elsewhere. Denis Hayes writes of the educational standard of the men, the desire of the 'politicals' to obstruct anything they could. There are many reports from all kinds of sources that the Army was generally more sympathetic towards conscientious objectors than civilians and especially civilian women and also that there was no opportunity for promotion within the NCC.(19) As for the more tendentious parts of the article, none of the events described is unbelievable; it would be much more unbelievable if the contention was made that 'incidents' never occurred. The question is really whether they were the exception or the rule. Hayes does not give any prolonged mentions of incidents in his book so it can be assumed that they were the exception. He puts more stress on the worries of non-combatants that the work they were doing, innocuous enough by itself, might have had an 'ultimate purpose' of which they were unaware and could only suspect. He also points out the frustration that so many well-educated men must have felt at the apparent futility of much of their work. Attempts to relieve this frustration took the form of 'evening classes in modern languages, orchestra, gramaphone recitals and play-acting', frustration which these activities, Hayes believes, 'failed to remove'.(20)

In 1940 a Company of the NCC attempted to produce a magazine for non-combatants, named 'Bless 'Em All: the Chronicle of No. 1 Company NCC'. However, Major-General Sir Alfred Knox and some other Conservative MPs took exception to an article in it on Armistice Day and Knox asked a question about it in the House of Commons. In his reply Captain Margesson, Eden's successor at the War Office, explained (21)

> When the first issue appeared, it was found that certain of its contents and in particular an article on Armistice Day, were of an objectionable nature. The distribution of the first issue was therefore stopped, and no further issues will be published.

Another way of occupying the time might have been by evangelising their faith either in religion or pacifism or both. A CBCO reporter wrote to headquarters about an applicant who expressed a desire in his Tribunal to 'carry on constructive peace work and witness and who said that he felt it would be his duty to continue to expound the pacifist ideal and teaching as well as to render some helpful service to the community.' The reporter was worried because the Tribunal managed to persuade the applicant to perform non-combatant duties. The Tribunal 'stated emphatically that he would be perfectly free to expound his pacifist principles to other members of such a unit.' But surely this would be regarded as sedition in the Armed Forces, wrote the reporter.(22) However, it did not apparently matter if a non-combatant preached, as long as he was preaching to the converted. If, as with Company no. 1, there was an attempt to print pacifist propaganda material, it could easily be prevented. Generally officers and soldiers from other corps tolerated objectors' views and even, on occasion, found that they could admire and respect them.

In fact, many non-combatants displayed an exceptional willingness to do as much as possible to help the Armed Forces prosecute the war. Altogether approximately 2,400 out of nearly 7,000 non-combatants transferred from the NCC to other units or other work without renouncing their status as conscientious objectors. Some of this number may be accounted for by the pressure brought to bear on them to do so; generally, however, it was because of the frustration of belonging to a corps, the usefulness of which was always in question. (23) At first non-combatants were concerned that they might have to lose their non-combatant status to transfer to, for instance, the RAMC. But Anthony Eden had made it clear as early as November 1940 that, 'It would not be necessary for a man to renounce a conscientious objection to combatant duty in order to be transferred to the RAMC.'(24) This was true of all other transfers as well. Once these fears were allayed therefore, non-combatants transferred at will. Hayes calculates that 216 men transferred to the RAMC (which, during the war, developed a shortage of manpower), 465 to the Royal Engineers where they could perform the dangerous work of bomb disposal, 607 to the 'smoke companies' work involving laying smoke-screens to inhibit enemy attacks, 400 to the RAPC where they could help in the administration and organisation of prisoners of war, 162 to the Paratroopers as medical orderlies and 547 to coal-mining when the shortage of coal getters became so serious later in the war.(25)

Despite the obvious dangers of bomb disposal, perhaps the non-combatants attached to the Paratroopers as medical orderlies had the most dangerous, exciting and fulfilling work. Some of their work was abroad and behind enemy lines and they had a chance of demonstrating in the most positive way the futility of war by tending not only wounded Allied soldiers but those of the enemy as well. The 'Sunday Graphic' and 'Sunday News' both carried the story of a German officer captured by the Allies in France. The officer explained that he had been shooting at British Paratroopers 'and what happened when I found my first Englishman is the reason why I say you people are mad. I lifted my revolver and fired at him twice.' The newspaper continued the story: the shots missed, the paratrooper dodged behind a tree and instead of firing back, to the amazement of the officer, he cried out in German,(26)

> 'Tell me Herr Officer, have you any blankets I can borrow?' 'Who are you? What is this nonsense about?', asked the German lieutenant. 'I'm a C.O.', said the paratrooper calmly. 'Then', said the Nazi, 'Gott in Himmel, what are you doing here?' 'Oh', said the paratrooper, 'Our blankets dropped in the marsh and we've got some wounded men - a couple of Germans among them - in a cottage up the road and I'm looking around for something to keep them warm. Can you help me?'

Like the experience of ordinary soldiers, much of the non-combatants' work in the war was tedious, frustrating and very wasteful of their considerable talents which Army life failed to utilise. But many made stringent efforts to find work within the Armed Forces which, as long as it could be reconciled to the particular stand a non-combatant was making, was interesting, demanding and rewarding.

'CAT AND MOUSE'

Of all apparent injustices and unfairnesses which upset conscientious objectors and their sympathisers, the one that raised the most indignation was 'cat and mouse'. The phrase originated, of course, with the suffragettes. Repeated prosecutions for the same offence, stays in prison at the finish of which only awaited another prison sentence was reminiscent of the way in which a cat teases a mouse, capturing it, releasing it, and then capturing it again, an agonising vicious circle from which there is no escape. The term was also used to describe repeated court-martial sentences in the First World War. If a declared conscientious objector was forced to enter the Armed Forces, he would refuse the first order given him (usually to put on his uniform) and would be court-martialled and sentenced for varying lengths of time in a military or civilian prison. On his release he would again refuse the order and the whole process would be repeated. Hayes calculates that in World War I some 655 objectors were court-martialled twice, 521 three times, 319 four times, 50 five times, 'and an heroic 3 six times'.(27) Many people, including a number of MPs who vividly remembered the scandal of the last war, were at pains to ensure that it could never happen again. The perfect solution, they felt, was to legislate against it and much pressure was put upon the Government to include in the Military Training Act and the National Service Act 1939 some provision to prevent 'cat and mouse' in World War II.

On 18 May 1939, during the debate on the Military Training Bill, Fred Messer MP (Labour) introduced a new clause concerning the 'penalty for refusal on conscientious grounds to obey an order', the gist of which was that if a soldier or a non-combatant refused an order on conscientious grounds, he ought first to appear before a military tribunal. If this tribunal suspected that the refusal had been on conscientious grounds, the man ought to be referred to a civilian tribunal. Messer asked (28)

> Are they to be forcibly stripped, forcibly dressed, forcibly detained; and, after their period of detention, if they continue to refuse are they to be punished again for a second refusal?... Some of us remember the cat and mouse business during the last War ... there were men who were punished repeatedly and the very repetition of the punishment proved conclusively that they had a genuine conscientious objection.

Messer felt that his clause would help the situation but the Attorney-General, Sir Donald Somervell MP (Conservative) advised the House not to accept the clause. It would, he felt, mean an indirect way of getting a further appeal; it would force the court-martial to state that the Appellate Tribunal had been wrong, an untenable situation. In any case, although mistakes might be made, they could only be on a small scale. Dr Alfred Salter MP (Labour) pointed out that 'mistakes were made on a gigantic scale on the last occasion, and there is every reason to apprehend that they may be made again.' (29) But the Attorney-General persisted. If a second Appellate Tribunal was absolutely necessary then he said that the House would have to set one up but not in the manner suggested by the clause. Then Arthur Creech Jones MP (Labour) recalled at length his experience in the last war. (30)

> I actually served periods amounting to about 3 years and 6 months. All the time the cat and mouse rule operated as far as I was concerned. It was recognised all through this course that I was a perfectly genuine person. Nevertheless I had been caught up in the military machine and the cat and mouse arrangement began to operate.

H.B. Lees-Smith MP (Labour) and Campbell Stephen MP (ILP) both continued to press the Government to produce a positive answer to the question even if the new clause was not suitable. Apart from the sheer injustice of the matter, Stephen pointed out, 'it was not fair to the military' to have to deal with conscientious objectors. (31) This pragmatic view of 'cat and mouse' was shared by civil servants in the Ministry of Labour in 1941 when they wished that prisoners could be discharged straight away to the NCC, but they knew that under the National Service Act this was not possible. A civil servant commented in a memorandum on the subject 'The men concerned are only a nuisance to the Army and might be some use elsewhere.' (32) The Government, however, had been prepared to compromise over this issue in 1939. Leslie Hore-Belisha, Secretary of State for War, speaking on behalf of the Government, said he had every sympathy with the critics. While he felt that it was impossible for the Government to accept this particular clause, he gave an undertaking that the Government would take some action. (33) A new amendment was therefore introduced while the House of Lords considered the Bill. It indicated that if a man served a three-month court-martial sentence and claimed that his offence had been committed on conscientious grounds, he could then apply to the Appellate Tribunal. The Lord Chancellor reassured the House of Lords that the Government did not want a 'get-out' clause but, 'the Government gave a pledge that they would introduce a clause of this kind.' (34) The trouble with this arrangement, as far as conscientious objectors were concerned, was that there was no compulsion on the courts-martial to issue sentences of three months or more. If they issued sentences of less than this figure, even one day less, the objector would be subject to a repeated prosecution and the cat and mouse circle would begin.

Fears of this eventuality were realised. Courts-martial frequently gave sentences just short of three months. Whether this was because the courts-martial were purposely defying the intentions of the Government or because the offences committed were too trivial to be given sentences of three months is a matter for speculation. The 'New Statesman' was convinced that the courts-martial were purposely giving 'short sentences'. In July 1941 it said that 'In the immense army machine, it is possible for minor officers to contravene the orders of the Government and the War Office.' (35) The Government, on the other hand, took the opposite view. In 1940 Oliver Stanley,

Secretary of State for War, in answer to a question about 'short sentences' replied,(36)

> I am not in a position ... to make or order to courts-martial that they were to give sentences which they did not consider to be in consonance with the gravity of a particular offence with which a man was charged.

Not only were courts-martial giving 'short sentences' but they were also giving sentences of detention rather than imprisonment. A period of military detention did not qualify a man for a hearing at the Appellate Tribunal, however long that sentence might be. The 'New Statesman' commented, 'The War Office has sent full and clear instructions for Presidents of Court-Martial to the Commands; those instructions appear to have been disregarded.'(37) A month after the publication of the article, Captain Margesson, now Secretary of State for War, wrote to the 'New Leader' explaining that, while the Government agreed that 'cat and mouse' was undesirable, it could only inform Presidents of courts-martial that three-months' imprisonment was the qualifying sentence. There was no obligation on the courts-martial to award this qualifying sentence.(38) The first person in the war to suffer from this 'loophole' was Kenneth Makin, a man whose application had been turned down by the Appellate Tribunal and who had been ordered to take up non-combatant duties in the Armed Forces. Makin had agreed to a medical examination assuming that it was not the right juncture to make his stand, and on entering the RAMC had refused to put on his uniform. At the court-martial he was sentenced not only to less than three months but to detention rather than imprisonment. On both counts, therefore, he was not entitled to a hearing from the Appellate Tribunal and it looked as though he was to be the first 'cat and mouse' victim. However, the case was given so much publicity, especially in Parliament where John McGovern MP (ILP) brought up the case and invoked the sympathy of many MPs that Makin was eventually allowed to appear before the Scottish Appellate Tribunal where he was exempted from military service on condition that he undertook work in agriculture.(39)

But the fight for court-martial sentences of three months or over and sentences of imprisonment rather than detention went on. After a great deal of pressure from the CBCO and from certain MPs, the War Office confirmed in a letter to all Commands in May 1940 that (40)

> a sentence of detention ... does not entitle a soldier to exercise his right of appeal to the Appellate Tribunal ... or make him eligible to appeal before the Advisory Tribunal .... I am therefore to suggest that you will consider the advisability of bringing this point to the notice of the Presidents of Court-Martial assembled for the purpose of trying cases of this nature.

Apart from conscientious objectors who had entered the Armed Forces unwillingly, there were soldiers who had never registered as conscientious objectors and who had never appeared before a Tribunal but who developed conscientious objections to part or all of their work in the Armed Forces. These men were not provided for in Section 13 of the National Service Act 1939. While legally this remained true throughout the war, the Secretary of State for War made a concession in May 1940 which in practice allowed these objectors the same rights granted to those who had registered in the provisional register of conscientious objectors. The only difference was one of nomenclature: after three months' imprisonment they were entitled to a hearing of the Advisory Tribunal, which was simply the Appellate Tribunal sitting in an advisory capacity to assist the War Office in determining when a soldier could be discharged from the army.(41) By 1941 remission of sentence was allowable in that soldiers who were recommended for discharge from the Armed Forces either by an Advisory or Appellate Tribunal were immediately discharged and had the rest of their sentence remitted.(42)

All this assumes that men who were allowed hearings at the Tribunals convinced the Tribunals that they were genuine conscientious objectors. But this was by no means true of all cases. At the end of 1942 a great deal of publicity was given to the cases of Gerald Henderson and Stanley Hilton, Jehovah's Witnesses who had been repeatedly imprisoned. The Minister of Labour was reluctant to recommend their discharge from the Armed Forces because the Appellate Tribunal 'has repeatedly held that the men are not conscientious objectors and therefore it would be undermining the integrity of the Appellate Tribunal to deal with them in any other way.'(43) However, by 1943 an administrative concession had been made that, on completion of a third sentence, a man would normally be discharged, 'services no longer required'.(44)

At last the circle of 'cat and mouse' had an end to it. It meant that no man, whether a genuine conscientious objector or not, would have to spend more than three terms in prison if he claimed that the offence was committed on conscientious grounds. Hayes's figures for those who were court-martialled for a refusal of an order apparently on conscientious grounds were calculated up to the end of 1946. By that date 1,050 men had been court-martialled, of which 636 had originally registered as consicentious objectors and 415 had reached a conscientious objection while in the Forces. Some 716 were court-martialled once, 210 twice and 10 three times. Before the administrative concession by the War Office 15 people were court-martialled four times, 2 five times and 1, Gilbert Lane, six times.(45) Up to the end of 1948, 808 men appealed to the Appellate Tribunals sitting in an Advisory capacity to be released from HM Forces, of which only 221 were refused.(46)

Although the term 'cat and mouse' is usually confined to a description of repeated prosecutions in the Armed Forces, there was one other situation which invited the same treatment. It concerned the penalties for refusing the medical examination for entry into the Forces. Until 1941 the question in this context did not arise because those who refused the examination were 'indefinitely detained'.(47) However, in 1941 the National Service Act made the penalty for refusing the court's order to undertake a medical examination a maximum of two years' imprisonment or £100 fine. There were worries that at the end of a long term of imprisonment objectors should be released only to be reprosecuted for the same offence. As a response to the mounting pressure from the CBCO and sympathetic MPs, the National Service (no. 2) Act 1941 included clauses designed to prevent 'cat and mouse'. Ernest Bevin explained in Parliament the need for these clauses.(48)

> Clause 5 seeks to remedy an anomaly in the treatment of Conscientious Objectors. In the 1941 Act, we introduced an amendment

> regarding medical examination. Where a man refused medical examination he was brought before a court and could be sentenced to imprisonment ... the man who is called up for a medical examination and received three months for refusal may then be subject to a sort of cat and mouse procedure. We have not yet exercised that procedure, but I think the House will agree that it is objectionable. Under this Bill, we have put a man who refuses medical examination in the same position as if he had been in the Army and had been court-martialled.

Again, similarly to the army situation, this new law did not preclude the possibility of men being rejected by the Appellate Tribunal and being summoned again for a medical examination. But in 1942 an administrative decision was taken and a prosecutor for the Ministry of Labour announced in court that in future conscientious objectors who had been sent to prison for three months or more for refusing to submit to a medical examination would not normally be prosecuted again for the same offence.(49) At last 'cat and mouse' seemed to be a thing of the past.

CONSCIENTIOUS OBJECTORS IN THE ARMED FORCES

Apart from 'cat and mouse' there were other questions which concerned conscientious objectors in the Armed Forces. One of the most serious of these was what would happen to a member of the Forces if he declared a conscientious objection while he was abroad? Would he be court-martialled and imprisoned abroad? The War Office agreed with the CBCO in 1942 that Section 13 of the National Service Act did not cover offences committed abroad, but it thought it very unlikely that such a man would have to serve a term of imprisonment overseas.(50) In 1943 a case occurred in which an aircraftman in India declared a conscientious objection and was sent home as soon as was possible and allowed to appear before an Advisory Tribunal. Discharge was recommended and he was put on land work. Similar events occurred in another case.(51) It was fairly clear that, while no legal provisions for the protection of conscientious objectors abroad existed, the Services were in all cases sending the men home so that they had a chance of presenting their cases before the Advisory Tribunal. But no formal notice of this administrative step was ever made during the war and the CBCO was still investigating this procedure in 1961!(52)

In most cases the servicemen who 'changed their minds' with regard to their role in the Armed Forces were not officers. However, the taxing problem of what should happen to an officer who declared a conscientious objection was raised in Parliament when Rear-Admiral Beamish MP (Conservative) asked the Secretary of State for War what decision had been reached on the question of officers who had declared a conscientious objection to combatant service. Oliver Stanley replied that 'such officers shall be called upon to resign their commissions.'(53) Three cases had been recorded by February 1940.(54) Another factor which disturbed MPs was the delay between arrest and court-martial, not just for conscientious objectors but for any offender. In May 1941 Sir Richard Meller MP (Conservative) asked Captain Margesson about this matter. Margesson replied that he was aware of the problem and would try to do something about it.(55) Again in February 1942 Frederick Bellenger MP (Labour) asked about the delay. Sir James Grigg, the new Secretary of State, promised that he would try to get the delay shortened.(56)

Once at the court-martial the problem then arose of whether objectors who had refused to put on their uniform should be forced into their uniform for the purposes of the court-martial. In 1941 the CBCO decided that there was no warrant in military law for the contention that the accused before a court-martial must be paraded in uniform. The matter had been brought to their attention because the President of a court-martial at the AMPC training centre at Dingle Vale School, Liverpool had said, according to a witness, 'There are certain of you appearing in civilian clothes. You will have to be paraded in uniform whether you like it or not. Those people not in uniform will have to get dressed before they appear before the Court.'(57) In 1943 Major Cunningham of the War Office wrote to the CBCO asking it to 'instruct your members that once they have been placed on a charge of disobeying an order to wear uniform, they should then agree to wear uniform for the purposes of the trial.' The CBCO decided that, while they would give information as to the likely course of events if objectors refused to wear uniform, they would advise objectors to follow their own conscience on the matter.(58) In fact there is little evidence that objectors appearing before courts-martial in uniform had been 'forcibly dressed'. The ignominy of 'forcible dressing' was more common when an objector arrived in the Armed Forces. If he refused to put on uniform, he might be physically assaulted and forced into it. In this way he could be prevented from exercising his right to a court-martial hearing because he had apparently agreed to the order, and it could not be proved that he had refused.(59) But it is clear either that objectors found they could, in all conscience, wear a uniform just for the trial, or that the Armed Forces did not insist upon it.

At the court-martial the accused was allowed to be accompanied and represented by a friend. But apparently some court-martials refused to allow the friend to be present. Cecil Wilson MP (Labour) asked in Parliament in 1941 if the friend of an accused could ever be kept out. Duncan Sandys MP, Financial Secretary of the War Office, answered that (60)

> The friend of an accused is entitled to be present in court at all times during which the accused himself is before the court and desires his friend to be present.
>
> Wilson: If I send the hon. Minister cases where this right has been refused, will he look into them?
>
> Sandys: Certainly, Sir.

The question was never raised again and presumably no more 'friends' were withheld entry into a court-martial.

From the time of arrest until the end of whatever sentence was imposed by court-martial, unless the sentence was to be served in a civilian prison, conscientious objectors had time to examine the conditions in military prisons and detention barracks. In fact not many sentences were served in military prisons; where the sentence was one of imprisonment it was usually served in a civilian prison. In April 1941 Cecil Wilson asked for the numbers of men presently confined in military prisons and detention barracks. On 1 March 1940 there had only been 1 conscientious objector, on 1 June 1940 there were 2, on 1 September there were 11, on 1 December there were 17, and on 1 March 1941 there were 3.(61) The reason for the sharp drop in numbers in 1941 was that by that time HM Prisons were not so overcrowded and most sentences would be served in civilian prisons. The precedent for this was set in the First World War when an Army Order in Council instructed that all sentences of men who claimed that their offence was committed on the grounds of conscience should be served in civilian prisons.(62) No such order was made in the Second World War but in practice it was largely so.

However, for some, many hours were spent in military prisons and especially in detention barracks. In March 1940 John McGovern MP (ILP) suspected that in some cases physical pressure was being brought to bear on offenders either waiting for their court-martial or after their offence was known. 'May I ask the Minister ... to see that at least there should be no brutality used on these men, as there have been reports of brutal conduct towards them?'(63) John McGovern may well have been right in his suspicions since it was only six months later that the scandal at Dingle Vale School, an army training centre in Liverpool, was revealed to the public. A small number of conscientious objectors who had been told to take up non-combatant duties and had refused orders, arrived there to await court-martial. The events at Dingle Vale, and a similar incident at the Old College, Liverpool, were the only occasions, according to Joseph Brayshaw who wrote a small piece on it for Denis Hayes's book, of 'organised savagery directed expressly at conscientious objectors during the War.'(64)

The first question asked in Parliament about events at Dingle Vale was on 17 October 1940. Edmund Harvey told the Secretary of State for War that there had been 'complaints of a serious nature as to the treatment of a number of conscientious objectors attached to the AMPC in a North-Western district.... What action has been taken to deal with the situation?' Sir James Grigg answered that he had heard complaints and that there would be an urgent inquiry into the matter. He went on, 'I need hardly add that it is the desire of the Army to treat conscientious objectors with scrupulous fairness in whatever unit they may have been called upon to serve.'(65) A few days later Glenvil Hall was describing in more detail the nature of the complaints about events in Dingle Vale. Many conscientious objectors

> have been treated in a brutal manner, being kicked, beaten with rifle butts, placed on bread and water in solitary confinement, dragged from their cells, marched round a square and prodded on with rifles; that the colonel in charge knew of this conduct but refused to intervene; that he insulted the men when they asked for a court-martial.

Sir James promised that a court of inquiry would be held to investigate allegations regarding the treatment of conscientious objectors and that he would tell the House the result of the inquiry as soon as he himself was acquainted with it.(66) At the beginning of November several MPs asked Grigg whether the inquiry had been held in public and what were its findings. Grigg said that it had met in private and that it was now preparing its reports. Glenvil Hall again suggested in strong terms that the inquiry should have been held in public. Grigg said, 'I do not think there is any reason to direct any suspicion against the court of inquiry. It consisted of very highly-placed and respectable officers.' Major Milner MP (Labour) asked if the Minister would see that a list of witnesses was published.(67)

> It has been said that some of those who might have given evidence may have been spirited away - with what truth I do not know.
>
> Grigg: I will look into that point.

On four occasions after this MPs asked for news of the progress of the report, but on each occasion Ministers from the War Office told them that the report was subject to delay because of the wealth of evidence which was apparently 'voluminous'.(68) At last on 28 January 1941 Captain Margesson announced,(69)

> The report was received on 28 November last. After careful consideration I have decided that one officer and six non-commissioned officers, against whom allegations have been made, should be tried by court-martial, and instructions have been issued accordingly. It will be appreciated, therefore, that I could not properly make any further statement at this stage.

On February 1941 James Maxton asked on behalf of John McGovern whether the Secretary of State for War intended to publish the report of events at the Dingle Vale Camp. Margesson replied that he could not make any comment as the case was under court-martial. (70) Then McGovern himself asked the reason for the failure to have the Commanding Officer of the training centre court-martialled for brutality. Margesson replied, 'The proceedings of the court of inquiry did not disclose any facts on which such a charge could be made against the commanding officer.'(71) On 6 May 1941 Benson asked whether, as a result of the court-martial (at which there had been a conviction of two non-commissioned officers and one officer) the Secretary of State for War would reissue the War Office letter of September 1916, forbidding such coercion and requiring offenders to be remanded for court-martial in accordance with military law. Margesson answered that he would consider it.(72) The letter was not reissued. Because the proceedings of the court of inquiry were in private it is not clear on what evidence the officers were court-martialled. However, A. Joseph Brayshaw sat throughout the court-martial of the officers and describes the events in Hayes's book. Brayshaw maintains that the prosecution was inept and the defence untruthful and that the trials were 'a pathetic travesty of justice'. (73) In any case the sentences given were merely severe reprimands and one sergeant was demoted to the rank of corporal. Whether or not the trials were fair and whether or not the sentences were adequate, at least the occurrence of the court-martials and the sentences acted as a deterrent for the rest of the war. No similar events were recorded again.

The physical conditions of the detention barracks were brought into question in Parliament at the beginning of 1942. In answer to a question from Frederick Bellenger MP (Labour), Duncan Sandys said that there had been a departmental inquiry into the state of the detention barracks and it showed that the conditions were 'very good'. Certain recommendations had been made, however. Bellenger asked if the report was to be published. Sandys replied that he would consider the matter.(74) In May of the same year John Banfield MP (Labour), who had apparently visited some detention barracks, maintained that conditions were bad, that they were insanitary and worse than civilian prisons. Sydney Silverman MP (Labour) asked if the report was going to be published and Sandys answered that it was a Departmental report and that he was not prepared to publish it.(75) There was, then, a considerable difference of opinion between certain MPs who had visited barracks and the civil servants who wrote the report and the Minister who accepted the findings of it. Probably as in civilian prisons, conditions varied.

THE MEDICAL EXAMINATION AND PENALTIES FOR REFUSAL

For those whose application had been rejected, or who had been given non-combatant duties but felt that they could not undertake them for reasons of conscience, there were three courses of action open. They could bow to the decision of the Tribunals and enter the Armed Forces, they could refuse to be medically examined (a prerequisite for entry into the Armed Forces) or they could, once in the Armed Forces, refuse an order, say to put on uniform. The last choice has already been discussed although it should be added that some maintained that objectors who allowed themsleves to be medically examined and then declared a conscientious objection in the Services did so under a misapprehension of the purpose and function of the medical examination. They simply did not realise that refusing to be medically examined was a sure way of avoiding entry into the Armed Forces, however unpleasant the consequences might have been. However, it is possible that some objectors felt themselves to be martyrs for the 'cause' and entered the Armed Forces with the express intention of demonstrating to officers and soldiers alike that there was resistance to warfare within the ranks and thus of disturbing the Army machine as much as possible. Not all could have been innocent victims of a misunderstanding; there were those who, rightly or wrongly, felt it their duty to make their stand from within the Armed Forces.

Most, however, who were determined to continue with their stand after their rejection by the Tribunal, settled for refusing the medical examination and awaiting the consequences. At the beginning of the war the only penalty for refusing medical examination was a maximum fine of £5 or one month's imprisonment in default. In addition a court could order a man's arrest and detention in order to 'secure his attendance before a medical board or consultant examiner'.(76) There was no maximum limit on this period of detention and soon it became known as 'indefinite detention'. This was altogether an undesirable development and as a result of pressure from the CBCO and from MPs and other individuals, the Home Office issued a memorandom to all Justices' Clerks in the summer of 1940. This instructed that the maximum periods of detention should always be specified.

Despite this development the authorities were dissatisfied with the arrangement and felt that the only answer was to increase the maximum sentence of imprisonment that could be given for refusal to submit to a medical examination. This was intended to deter not so much conscientious objectors but other conscripts who were taking advantage of the small penalties involved to avoid entry into the Armed Forces. In the National Service Act 1941 the courts were allowed to order a man to submit himself to a medical examination. If he did not comply with this order he was liable to a maximum period of two years' imprisonment or a fine of £100 or both. Forcible examination was ruled out at this time because of, as Ernest Bevin put it, the 'repugnance' of the idea.(77) But the difficulties for conscientious objectors were not solved because, despite pressure put upon the Government in the Commons and the Lords to write into the Act some provision for the prevention of 'cat and mouse', no such provision was made.

A campaign was launched and supported by many sections of the community to persuade the Government to find a way to avoid 'cat and mouse', a phrase by now familiar to all who read newspapers. The campaign was successful because, as has been explained, the National Service (no. 2) Act was passed at the end of 1941. This allowed the procedure for objectors refusing their medical examination to be exactly similar to that for those declaring a conscientious objection within the Armed Forces. For those who failed to convince the Appellate Tribunal a new measure was instituted at the end of 1942 assuring them that they would not normally be prosecuted again for the same offence.

It now remains to discuss the experience of objectors in civilian prisons remembering that it was not only that of those objectors who had refused their medical examination but also of those who had refused an order in the Armed Forces and who were serving their court-martial sentence in a civilian prison. Perhaps the most distressing part about serving time in prison for any sensitive person was neither the physical conditions, nor the attitude of prison officers nor the attitude of other prisoners, but the social stigma that it carried outside prison. Some tolerant and understanding friends could no doubt support the conscientious objector in prison but for many others it must have been shocking and humiliating that an otherwise gentle, respectable and law-abiding relation or friend should find himself alongside murderers, thieves and other offenders. Being released from prison was especially difficult for conscientious objectors because they were often rejected by those they loved as a result of the stand they had made, and, of course, the possibility of obtaining work again in whatever their profession might have been was vastly reduced by having a prison record. Almost all conscientious objectors with a prison record whom the author spoke to experienced this. This sort of suffering is impossible to quantify, unlike physical discomfort in the prisons, but nevertheless it needs to be understood to gain an accurate impression of the experience of those objectors who spent time in civilian or military prisons.

Physical and mental discomfort in prisons was, however, also experienced. Sybil Morrison, the historian of the PPU, spent the Battle of Britain in Holloway Prison serving a sentence for using 'insulting words and behaviour' while addressing crowds in London on pacifism.

> The Governor told me I had committed a worse offence than any other prisoner, and the officers frequently told me that if there was an invasion I at least would never be released. I knew a chill of helplessness and fear that I hope not to experience again.

She recorded that a decision was eventually made to unlock women's cells during bombardment, although this could not safely be done for men. 'When a bomb sliced a wing off Walton Prison in Liverpool, one conscientious objector was among those who died.'(78)

It is clear that conditions in civilian prisons varied enormously from one place to another and from one period of the war to another. A good example is that of Wormwood Scrubs in London where for two weeks in February 1942 after prisoners had been transferred from Wandsworth, no visiting was permitted, the prison organisation was chaotic and even the most uncomplaining of conscientious objectors reported afterwards that the physical conditions were 'appalling', especially the sanitary arrangements and the quality of the food and the lack of it.(79) But in the following month prisoners were telling Quaker Ministers who visited that the food had vastly improved and that the warders' understanding of conscientious objectors had increased considerably. 'The Scrubs' was now being used as a central prison for all London objectors and temporarily for other objectors who had to be brought from provincial prisons to go before the Appellate Tribunal. Quaker Ministers who wrote frequent reports for the Friends Home Service Committee based in Birmingham found it interesting to hear accounts from objectors of conditions at other prisons compared with 'the Scrubs'. They placed the prisons in 'order of pleasantries'. First came Lewes, then Dorchester, then Maidstone whose Governor was particularly understanding. Fourth came Wormwood Scrubs itself followed by Exeter, Stafford and last Birmingham. By June 1942 Ministers reported that

> Discipline conditions have become very easy, and there is a good atmosphere and easy relations between prisoners and staff, and C.O.'s and their fellow prisoners. After three months dinner and supper is served at tables seating 12 men and free conversation is allowed.

In June 1943, at the time when Michael Tippett, the composer, was serving his sentence, a Minister noted that, 'A new improvement in conditions is that flowers in vases and pot plants are allowed on dining room tables.... I am now taking in a large bunch of flowers each Thursday morning.'(80)

Surprisingly there are few reports of any victimisation by prison officers or by fellow prisoners, both of which groups were equally puzzled by the presence of these men. On the whole objectors made 'star prisoners' as one warder put it.(81) Even attempts at evangelism of their views were apparently tolerated but the main evangelism was by example rather than by harangues about the righteousness of the conscientious objectors' position. In some prisons, Wandsworth, Strangeways and Wakefield, regular Quaker meetings of worship were being held in 1941 under the care of the visiting Quaker Minister.(82)

Perhaps the only good that accrued from the presence of conscientious objectors in HM Prisons during the war was the interest shown by those who represented them and sympathised with them about conditions in prisons. When questions were asked in the House they were mainly inspired by concern about objectors but any improvement made would obviously benefit all prisoners. With one exception,(83) at no time did any organisation ask for any preferential treatment for objector prisoners. In 1941 Rhys Davies MP (Labour) asked the Home Secretary if he was aware that prisoners were confined to cells for as long as 18¼ hours per day. Morrison replied that the war made the organisation of prisons difficult but he hoped to improve the situation. Davies made a special mention of Leicester Prison and he asked the Minister 'to consider whether it is possible to remove what are regarded by most sensible people as inhuman conditions.' Morrison replied that he would consider it.(84) Edmund Harvey asked a few months later whether prisoners were allowed to read paper backs. Reading was one way of using the time spent in prison to advantage; objectors had the opportunity to read solidly for months on end and became acquainted with books and authors which, without their stay in prison, they would never have encountered.(85) Morrison answered 'having regard to the excellence of many of these publications and of their advantage from the point of view of economy, I have instructed prison governors that for the present they may receive such books for prisoners.'(86) Prison diet was also discussed.(87)

The only really serious question mark which hovered over conditions at HM Prisons was caused by the death of a conscientious objector soon after his release from prison. In June 1943 John McGovern asked the Home Secretary if he had investigated the death of Victor Walker,

> sentenced on 2 December 1941 to 12 months imprisonment for an offence on conscientious grounds, who died some weeks after the end of the sentence, and what treatment was given by the authorities after the receipt of his brother's letter expressing alarm on 20 April 1940 and for diagnosis of heart trouble on 18 May 1942.

Morrison answered that the investigation was still proceeding.(88) But as there were no more reports in 'Hansard' it is difficult to tell what the results of the investigation were. The fact that the CBCO did not take up the case implies that the Board was satisfied that there had been no neglect on behalf of the prison authorities. Obviously a stay in even the best equipped and appointed prison cannot help a heart condition; in that sense it could be said that prison life accelerated Walker's death. But whether the authorities should have moved Walker to the prison hospital earlier remains an open question. There is no evidence to show that any conscientious objectors were manhandled or physically victimised in any way. One can only assume that the extract of the 'Diary of an Imprisoned C.O.' containing references to 'bruised bodies' was more literary licence than actual fact. The 'New Leader' printed the article in 1941:(89)

> Human beings, because of their convictions, are fighting alone against the power of the state. Whatever the state may be its weapons are always the same. Dull, mechanical brutality; degrading punishments, insufficient food and exercise; solitary confinement. Sometimes the prisoners do not break down. Sometimes from suffering they find super-human strength and from out of their bruised bodies and tortured minds there springs a dignity and grandeur which is of the essence of human freedom.... And then, somewhere, another lamp is lit.

The only preferential treatment for objectors which was asked was a quite natural request: that conscientious objectors should not be asked to perform war work in prisons. In the First World War a general instruction had been issued to HM Prisons that objectors were to be employed on mailbags while other prisoners were performing war work.(90) As there was no further mention of the matter in CBCO records it must be assumed that such an instruction was given.

Conscientious objectors, then, usually managed to survive their sentences in prison with good humour and courage. Although they were rarely victimised or persecuted within the prisons, they suffered for years afterwards the stigma of having a prison record. But perhaps the worst aspect of these prison sentences was the sheer waste of manpower and talent which most were only too willing to use in civilian life. This view is summed up in a letter to the 'Evening Express' in Aberdeen, a letter which was signed simply 'Common Sense':(91)

> Some people may be glad to see conscientious objectors being thrown into prison for their conscientious scruples, but to most thinking people it must cause a pang. Is it not for freedom of conscience that we are fighting at this minute.... Let the conscientious objectors have a chance to go on with some useful or civil work for their country ... instead of confining them at the country's expense when all hands are needed.

Chapter 7

THE PEACE PLEDGE UNION AND THE CENTRAL BOARD FOR CONSCIENTIOUS OBJECTORS

THE PEACE PLEDGE UNION

Before the Second World War, 112,905 persons had signed a pledge stating that they would renounce war and would refuse to sanction or support another and thus became members of the Peace Pledge Union.(1) The results of the Peace Ballot organised by the League of Nations Union had shown that, although most Britons would approve sanctions by arms if necessary, millions of them wanted lasting peace at almost any price. Even the overwhelming victory for the motion 'that this House will in no circumstances fight for its King and Country' at the Oxford Union Society, a debate which was undoubtedly given more significance at the time than it warranted, seemed to demonstrate that the youth of Britain had rejected war as a means of settling international disputes. The policy of appeasement which Chamberlain's Government pursued symbolised the anxiety and fear of everyone that another war would bring the carnage and misery suffered in 1914-18. The 1930s were the heyday of popular pacifism.

But by 1939 most British citizens accepted that war was inevitable and necessary in order to resist the growing pretensions of Hitler's Germany. The introduction of conscription, widely opposed in 1916, had been received calmly and with resignation. The PPU, however, made vigorous efforts at the outbreak of war to maintain its stand and to convince its membership that pacifism was not a doctrine for peacetime only but that it was just as pertinent, more so, in wartime. The chairman of the PPU, Stuart Morris, wrote in 'Peace News', the Union's weekly newspaper, a few days after war had been declared, that the advent of war had not proved the PPU's policy wrong. On the contrary, it had proved the non-pacifists' policy wrong. He now envisaged the PPU's role as maintaining the Pledge and supporting conscientious objectors, giving them financial aid, organising pacifist service corps and starting to make the greatest possible effort to find peace terms.(2) This campaign was successful in that membership reached a peak in April 1940 of 136,000.(3)

The PPU had begun to attract hostility as a result of its propaganda against the prosecution of the war. In 1939 it launched a

'Stop the War' campaign, thus inciting the indignation and anger of many people. In February 1940 Sir William Davison MP (Conservative) asked the Home Secretary, Sir John Anderson,

> whether his attention had been called to a recent statement by a judge who is Chairman of the East and West Riding Conscientious Objectors' Tribunal [Judge Stewart] as to the subversive activities of a body known as the Peace Pledge Union, who are picketing Employment Exchanges and endeavouring to induce men to join their organisation and avoid military service by claiming to be conscientious objectors, for which purpose special instructional classes have been arranged to supply objectors with particulars of the conscientious objections they should submit to the tribunals and as to the replies they should give when questioned, and what action is being taken by the Government in the matter?

Anderson said that he was aware of the statement and that he felt that these activities were an abuse of the facilities allowed for conscientious objection in Britain. He made the assurance that the whole matter was being 'very carefully considered'. Reginald Sorenson MP (Labour) said that any condemnation of the propagation of pacifist principles would reflect very seriously on such persons as George Lansbury MP (Labour) (who had succeeded Dick Sheppard as President of the PPU in 1938) and on many Christian churches. Were these persons and institutions subversive? Anderson retorted that there had to be a distinction made between a 'view which is genuinely held' and methods of endeavour 'to induce people to take advantage of the opinions of others.'(4)

The atmosphere changed for the worse, however, as it did for all conscientious objectors, when the events of Dunkirk, the Fall of France, followed by the Battle of Britain hardened the public's heart both against Hitler and against those not whole-heartedly committed to the war effort. Well-known writers who were supporters of the PPU like Vera Brittain, John Middleton Murry and Ethel Mannin lost sales 'because of their support of a most unpopular cause'.(5) The PPU was now a 'tiny band on a tiny and vulnerable rock in the middle of a raging torrent'.(6) 'Peace News', the PPU's weekly newspaper, never missed an edition even when its printers and distributing wholesalers refused to print and circulate in 1940 for fear of their reputations, and even of the law. Eric Gill, the well-known Catholic pacifist active in the Catholic organisation PAX, printed a four-page issue on his own printing press, Hague and Gill's, in the emergency.(7)

It was in June 1940 that John Middleton Murry was asked to edit 'Peace News'. Murry had suffered a great deal for the unpopularity of his views; he couldn't get any of his work printed and was apparently beginning to wonder how he could support his family.(8) The PPU National Council offered him £7.00 for a three-day week to edit the newspaper which was a 'far cry from the sober, sedate Athenaeum', which he used to edit.(9) He arrived at the Finsbury Park office of the paper at a challenging time; the membership had dropped precipitously since the outbreak of hostilities with several of the sponsors renouncing their pacifism, of whom Bertrand Russell was the most well known. Murry, however, was not perturbed; he believed that they would have left eventually anyway, and that what was left was human material with which he could work. He, unlike many of the pacifists now leaving the Union, was not so concerned to stop the war as to plan for a new community to take over amidst the ruins of the old. While this view was still not wholly shared in the PPU, the leadership of it was now a strong one: Max Plowman, Eric Gill, Lawrence Housman, Charles Raven, Alex Wood, Wilfred Wellock and George M. Ll. Davies. Within eight months Murry was able to report to them that circulation of 'Peace News' had grown from 9,000 to 18,000, and that its financial situation was sound.(10)

Murry's editorship, with his weekly 'Observer' column, was autocratic. Often two of the four pages were his own work. Shaw had said that Murry was too good a writer to be a good editor: his biographer noted that, 'he was quite capable of denouncing the 'official' policy of the Union in the leading article of its 'official' paper, promulgating his own instead.'(11) 'Peace News', however, survived the war with a circulation of nearly 20,000.(12)

Meanwhile, there was still marked hostility towards the PPU as an organisation. Accusations that the PPU held 'mock tribunals' and classes to teach people how to put forward the case for a conscientious objection were frequently levelled. A journalist of the Brighton Gazette reported,(13)

> The young men I saw argued their cases with admirable aplomb and the frequency with which the same phrases reappeared in their statements led to the suspicion that they had been well coached beforehand - that there was, in fact, some outside organisation working on their behalf which was interested in seeing that they were kept out of the Army.

In March 1940 three Conservative MPs, E.W. Salt, Sir Patrick Hannon and Sir Smedley Crooke complained in Parliament that the PPU was holding classes in the Midlands designed to teach men how to present a case for conscientious objection before Tribunals.(14) The question here was whether these 'mock tribunals' were being held to help already declared conscientious objectors to present their cases in a clear and articulate manner or whether they were designed to create plausible grounds for an objection which it was hoped that the tribunals would accept as conscientious, but which was in reality merely an objection, not motivated by conscience, to entering the Armed Forces. If the latter case was true, it would call seriously into question the integrity of leading members of the PPU. However, the Government felt that there was no need for undue concern. While the Home Secretary confirmed that he was keeping a careful watch on these activities he pointed out that, 'Reports reaching me from every quarter go to show that the activities referred to are, in fact, having no serious effect, beyond making a certain number of people very angry.'(15)

Despite this assurance MPs continued to show their concern over the activities of the PPU. Sir Henry Morris-Jones MP (Liberal National) maintained in June 1940 that the PPU was threatening to 'infiltrate' Wales and was 'endeavouring to thwart and divert our war effort'. On this occasion Eleanor Rathbone MP (Independent) pointed out that the PPU was receiving more attention than it warranted. 'Does the Minister not consider that the fact that the proportion of people applying for registration as conscientious objectors has steadily gone down, and is now something like 1 per cent, show the inefficacy of this organisation?'(16) But there were limits to the tolerance of the Government towards the PPU. In May 1940 six leading members of the PPU (17) were summoned to Bow Street Police Court for alleged breaches of the Defence Regulations based on a poster which announced: 'This war will cease when men refuse to fight. What are YOU going to do about it?'(18) Next month the men were bound over and the poster was withdrawn.(19) The 'Daily Mail' reported in May that 'Police will keep a keen eye on 'Stop the War' propagandists distributing pamphlets to the 27's who register today. Every Labour Exchange will be under the observations of police officers.'(20) Complaints about the PPU continued to be aired in Parliament. In 1942 Captain Gammons MP (Conservative) asked the Home Secretary why 'Peace News'

> is not barred in view of statements made in the issue of 13 March 1942, that the raid on the Renault works in Paris was made to produce a momentary and false impression of activity and to keep up morale at home, that the war was precipitated by partisans in Poland and Czechoslovakia and as it is openly asking for subscriptions to carry on Pacifist propaganda?

Herbert Morrison, Anderson's successor at the Home Office, agreed that statements of that sort 'go beyond anything which can be regarded as the legitimate expression of Pacifist views' and assured MPs that the newspaper would continue to be watched.(21) In 1943 Sir Smedley Crooke complained that the PPU propaganda in Birmingham was proving detrimental to the war effort. Osbert Peake, Under-Secretary of State for the Home Office, made a reply which succinctly summarised the spirit of the Government's attitude to the PPU throughout the war years, 'My present information is that the results achieved by this minority body are insignificant and it would be a mistake to exaggerate its importance by official interference.'(22)

Worries that the PPU would influence large numbers of men and women to assume the stance of conscientious objection to the war were unfounded. As Miss Rathbone had said, the numbers of those registering as conscientious objectors dropped rapidly after the spring of 1940. The PPU changed its policy under the influence of Murry; it became less political and more visionary. Some conscientious objectors' views were represented by the PPU during the war, but the organisation moved away from militant opposition to the fighting of the war and towards the planning for the peace. Ceadel argues in his chapter on pacifism in the Second World War that pacifists had gradually realised the political impotence of their beliefs, and through the social application of their pacifism in welfare and land work, they had been drawn into 'sectarianism'.(23) The number of pledges did not fall dramatically: in fact they remained fairly buoyant, but the isolation of the PPU as a political body increased.

THE CENTRAL BOARD OF CONSCIENTIOUS OBJECTORS

Fenner Brockway, Chairman of the CBCO, wrote in his foreword to Denis Hayes's book, 'Challenge of Conscience', that in 1939 (24)

> The need was not an organisation _of_ Conscientious Objectors but an organisation _for_ Conscientious Objectors, a coordinating body which would link the existing bodies and provide the specialised service which those of their members who came within the scope of the conscription laws would require.

The number of 'existing bodies' was considerable. The Peace Pledge Union has been discussed but there were many other religious and humanitarian societies and associations which all held an interest in conscription and conscientious objection. The Society of Friends, the Quaker association, with its long history of conscientious objection, was obviously greatly involved. The Fellowship of Reconciliation, a Christian organisation which had done much to help objectors in the First World War, was nearly 1,000 strong in 1939.(25) Other interested organisations abounded: the Independent Labour Party which had always sympathised with conscientious objectors; the No-Conscription Fellowship, another body which had done much for objectors in World War I; the Women's Cooperative Guild; Pax, the pacifist Roman Catholic group; Woodcraft Folk, a youth organisation similar in activities to the Scouts and Guides but not connected to any particular religious sect; many denominational pacifist groups including the Methodist Peace Fellowship and the Anglican Pacifist Fellowship; all these bodies had an interest in the plight of conscientious objectors and sought affiliation with the CBCO. The Fellowship of Conscientious Objectors, formed at the beginning of the war and the only organisation which attempted to create a movement of objectors,(26) was affiliated but wished to remain independent of the CBCO. Interestingly, the National Council for Civil Liberties decided not to affiliate in March 1940, though for what reasons is not clear.(27) Thus in all there were seventeen affiliated organisations, most of which were religious in character the others being either humanitarian or political.

Before December 1939 the CBCO had been called the National Joint Advisory Bureau, but at that time it was agreed that the name should be changed and that there ought to be a statement of the general direction in which the Board's future work should proceed. At the Board's first meeting the following statement was presented and agreed upon:

1 Nationally
   - a advice and information.
   - b records and care of conscientious objectors.
     - i individual cases where there is a failure to get the correct exemption.
     - ii statistics
   - c victimisation, persecution and hardship.
     - i help in finding jobs.
     - ii maintenance support (in cooperation with local Boards).
     - iii parliamentary action.
   - d information bulletins etc.

2 Locally
   - a local advisory bureaux.
     - i keeping records.
     - ii supplying information to the National Bureau.
     - iii advising local conscientious objectors.
     - iv seeking help of National Bureau.

v setting up a CO fellowship of all local conscientious objectors which should express views locally and to National Body.

b to encourage regional organisations to link together all local bureaux.

It was decided at the same meeting that affiliated organisations should send two representatives to Board meetings.(28) At the next meeting it was decided that the CBCO would use PPU premises in Endsleigh Street but that it would remain entirely separate from the PPU as an organisation.(29) The Board met at least monthly although on some occasions they met twice a week. In 1941 an Executive Committee was formed at the first meeting of which it was decided to keep handwritten minutes and prepare six-weekly reports for Board meetings.(30)

In February 1940 the various positions within the organisation were being sorted out. Lord Ponsonby, a veteran pacifist of the 1920s and 1930s,(31) and T. Edmund Harvey MP had apparently declined to accept the Presidency of the Board. Lord Arnold, the socialist peer, was named as the next choice, and failing that other names mentioned were Dr Alfred Salter, Dr George Sutherland, Wilfred Littleby, Arthur Creech Jones MP and Lord Farringdon. Later in the month, however, Lord Arnold agreed to accept the Presidency.(32) Fenner Brockway became chairman, Stephen Thorne Honorary Secretary and Stuart Morris (also General Secretary of the PPU) Liaison Officer. The position of Honorary Treasurer was shared by Isaac Goss, Sir Hugh Robertson and Foresster-Paton.(33) Denis Hayes became Information Secretary in August 1941.(34) In that year it was decided that the CBCO should try to register as a war charity for financial reasons and the application was made in June 1941, but was rejected by the London County Council in September.(35)

The work of the CBCO is described in Fenner Brockway's foreword; he shows that its main work involved advice to conscientious objectors and this was provided by various means. At local level it could be given personally but much of the work was done by publishing a considerable quantity of printed material on matters such as how to register as a conscientious objector, how to appeal against a Local Tribunal decision, a statement of soldiers' rights at court-martials and even a pamphlet entitled 'On Being a Good Witness'. One of the most valuable achievements, however, was the publication of a series of broadsheets unravelling the maze of legislation which wartime had brought. As Brockway said, 'These broadsheets, which were entirely factual, were in most cases O.K.'d by the Departments concerned before publication and gained a considerable reputation for their accuracy and clarity.'(36) The titles of the broadsheets are self-explanatory; the following are just examples: 'Police Court Procedure' (1941), 'Enrolment for Civil Defence' (1942), 'Compulsory Fire-Watching At Your Place of Work' (1942), 'Conscription for War Work' (1942), 'Medical Examination' (1943) and numerous others.(37) These broadsheets were so helpful on laws relating to wartime activity that other organisations, not connected in any way with conscientious objectors, bought them for their members.(38)

The other main service the Board provided for conscientious objectors was that of keeping a vigilant watch on all proposed legislation. CBCO officers were in constant contact with Government

Departments and co-ordinated their action with the Parliamentary 'Exemptions Group'. This was a group of members from both Houses formed to protect the interests in Parliament of those who applied for exemption from military service for reasons of conscientious objection or hardship. They were largely, but not all, all former conscientious objectors themselves. The 'Bournemouth Daily Echo' sneered at the number of MPs who were 'avowed to conscientious pacifism'. It reported that, out of 615 MPs, 'not more than half a dozen' fell into this category: the 'ILP trio' of James Maxton, Campbell Stephen and John McGovern, the communist, William Gallacher and two or three other Labour members.(39)

A CBCO record which was made in 1942 suggested that the following members of the House of Commons and Lords attended Exemptions Group meetings: Lord Arnold, whose bad health often prevented him from attending, Rev. James Barr, a Christian socialist sponsor of the PPU, George Benson, who apparently rarely attended, R.J. Davies, Lord Farringdon who, though 'keen', was rarely able to attend, Mrs Agnes Hardie, T. Edmund Harvey, A.Creech Jones who also rarely attended, John McGovern, James Maxton, Dr Alfred Salter, Rev. R.W. Sorenson, Campbell Stephen, Frederick Messer, Cecil H. Wilson and W.G. Hall.(40) On the whole this system worked well; the 'C.O.'s Hansard', published weekly by the CBCO and then bound into a book, was full of questions raised by various MPs in the Exemptions Group.

While it is clear from 'Hansard' that the MPs in the Exemptions Group worked hard for the interests of conscientious objectors, the relationship between the MPs and the Group at least once deteriorated sharply. Cecil Wilson wrote to complain to the CBCO that the Board was putting forward questions to be asked in the House 'simply for the purposes of propoganda', and that MPs were being approached separately rather than as a group so that when a question was raised in the House by an MP, the others were not properly prepared on the facts of the issue to support it. Joe Brayshaw maintained at a meeting of the Executive Committee that Wilson himself was responsible for this lack of co-operation.(41) However, Stuart Morris wrote a conciliatory letter to Wilson explaining that it was the Board's desire that good relations be maintained. He suggested that a member of the Board should be present at every Exemptions Group meeting and that the Board would greatly appreciate the presence of a Group member at Board meetings.(42) Despite these misunderstandings, however, the Exemptions Group served conscientious objectors well, not only during Parliamentary debates, but also by their numerous deputations to Ministers of the various Departments on questions concerning conscientious objectors. Sympathetic MPs were invaluable allies for the CBCO.

Officers of the CBCO were also in contact with Government Departments, exchanging letters and telephone calls and arranging deputations in the same manner as the Exemptions Group. The work of the CBCO and the Exemptions Group is perhaps best illustrated by studying in detail how they dealt with several important developments during the war which directly affected conscientious objectors. The first was that of industrial conscription, a term given to the powers which the Minister of Labour, Ernest Bevin, took in May 1940 under an Order in Council under which he could 'direct any person in the United Kingdom to perform such services within the jurisdiction as might be specified in the direction, being services which in the opinion of the Minister the person directed was capable of performing.'(43) The term 'industrial conscription' was apt; men and women were required to register as for military service, and in due course they could be directed, or 'called up' to any work which the Minister considered suitable. There was no 'conscience clause' attached to the Order and the CBCO was immediately concerned that conscientious objectors might be directed to war work and have no right of appeal to it. The PPU at this time submitted to the CBCO five objections against industrial conscription. Firstly, it inhibited personal freedom, secondly, it 'cut across' a sense of vocation, thirdly, it increased the power of central Government, fourthly, it might 'push labour into private industry and exploitation' and lastly, it involved an uprooting from social and cultural backgrounds.(44) These might have all been valid criticisms of industrial conscription, but they had no more relevance to conscientious objectors than any other persons. Direction to war work was the possibility which concerned the CBCO most; additionally there was the problem of those who might conscientiously object to being directed to any sort of work on the grounds that it was made necessary and desirable by the war and was obviously designed to further the war effort.

The CBCO and the Exemptions Group's aim was first and foremost therefore to persuade the Minister of Labour that there should be a right of conscientious objection to direction. There was some time to present this case for the first registrations did not become necessary until March 1941. Meanwhile the officers of the CBCO had been pressing the Minister of Labour at least to allow conscientious objectors the opportunity to record their position on registering. As a result the Minister decided that he could allow the setting up of Appeal Boards to which objectors and anyone else with a grievance against their direction could appeal. The CBCO, however, wanted to be sure that the Appeals Boards would allow a conscientious objection to be a ground for exemption. The Minister wrote to the Board saying that he felt that there was no reason why the Appeal Boards should not take a conscientious objection into consideration. The Exemptions Group also took up the matter with the Minister, the CBCO tried to elicit support from the then Archbishop of York, Dr Temple, and the Society of Friends also engaged in correspondence with the Ministry. The Minister, however, promised all along that he would administer the Order in such a way that would make it unnecessary for any conscientious objector to feel persecuted or victimised in any way. He promised in particular that no objector would be directed to munitions work.(45)

By and large the Minister kept his promises to the CBCO and the Exemptions Group. The contents of a memorandum sent to all National Service Officers in 1942 show quite clearly that the Minister was determined to administer the Order as fairly as he could. While there was 'no legal right of conscientious objection to industrial work', the Minister had pledged not to force any objector to perform munitions work and he gave instructions that this pledge should be 'liberally interpreted'.

> The National Service Officer should also remember that conscientious objectors will often work hard at useful jobs if tactfully handled, and that the main object of the Department is to supply labour for the national effort. The result will not be achieved by the imprisonment of a conscientious objector who is doing, or is willing to do, work which would earn deferment in the ordinary way or which would secure non-withdrawal. There may even be cases where the conscientious objectors should be left undisturbed in work of less importance rather than risk complete loss of production.

The memorandum went on to outline what should happen to the different classes of conscientious objectors: unconditionally exempted objectors were available for civil employment but direction could be given to them only after reference to headquarters; conditionally registered objectors could only be given work which fell within the conditions laid down by the Tribunals and only when the objector was not complying with his conditions, or when he was complying but full use of his capacity to work was not being made; objectors registered for non-combatant duties were liable for direction; provisionally registered objectors were liable for direction but had to be allowed to attend their Tribunal hearing and the Minister felt that it might be easier to wait to give any direction until after the Tribunal's decision had been made; persons who maintained that they were conscientious objectors although they had not yet registered as such (because their age group had not yet been called upon to register) were liable for direction.(46)

The combined efforts of the CBCO and the Exemptions Group may not have been successful in forcing the Ministry to accept the right of conscientious objection to industrial conscription but in practice they had succeeded in making the Minister very wary of directing conscientious objectors against their will. Perhaps the reason for this success lay in the massive publicity given to this wartime measure, not just by the CBCO but also by all kinds of newspapers and other publications. Industrial conscription was one of the least popular of the powers taken by Ernest Bevin, especially among the Trade Unions from whose ranks Bevin had risen. Every part of it was thus administered with extra care and understanding and this benefited conscientious objectors as much as everyone else. But that it was a touchy issue was in no doubt; a civil servant writing in July 1942 commented in a memorandum to a colleague: 'The Department is not unnaturally anxious to be assured that what we propose to say on this controversial subject is strictly in accord with Ministerial policy.'(47)

Despite this Ministerial sensitivity to the problems posed by industrial direction, there were a number of prosecutions of conscientious objectors and especially a large number of women many of whom, while liable for direction, had not yet had an opportunity to declare a conscientious objection formally.(48)

Another group which proved adamant against direction was the Jehovah's Witnesses who, in accordance with their belief in Divine work, refused to acknowledge any direction by the State. In 1944 the CBCO was concerned that Jehovah's Witnesses were being unfairly and repeatedly prosecuted and an officer wrote to Dr Temple, now

Archbishop of Canterbury, asking him to make representations to the Minister of Labour on behalf of Jehovah's Witnesses. The Archbishop was at first reluctant to take this step. He wrote back saying that his conscience did not allow him to permit himself to plead on behalf of Jehovah's Witnesses. He felt that it did not seem reasonable that the State should be expected, in a great emergency, to accept an individual's claim that his action was dictated by conscience, or to accept the claim that conscience directed an individual to ignore any requirements made consequent upon the State while it was engaged in warfare.(49) Even so, after a meeting with Rev. Henry Carter and officers of the CBCO, the Archbishop decided to approach the Minister on the subject of 'third prosecutions with special reference to Jehovah's Witnesses'.(50) Altogether there were 610 prosecutions of conscientious objectors on charges relating to industrial direction up to 1948,(51) a very small number in comparison with prosecutions relating to compulsory Civil Defence duties. The credit for the relatively small number of prosecutions should go to Ernest Bevin for his tactful handling of the administration of the Order in relation to conscientious objectors, but it is also quite clear that the Minister's actions were often a direct result of the many representations made to him by the CBCO and the Exemptions Group.

A second most complicated and vexing issue affecting conscientious objectors which arose during the war was that of compulsory fire-watching duties. Until the beginning of 1941 fire watching had been completely voluntary. Employers would make arrangements with their staff for guarding places of work at night, and domestic premises were left to the owners to protect. In January 1941, however, the Minister of Home Security, Herbert Morrison, issued two Orders, the Fire Prevention (Business Premises) Order 1941, and the Civil Defence Duties (Compulsory Enrolment) Order 1941, the latter designed to protect residential areas. Tribunals to deal with exemptions from these Orders were set up by the Civil Defence Duties (Exemption Tribunals) Order, 1941. A number of persons could be exempted from these duties: those who were already engaged in other wartime activities like the Home Guard and Civil Defence, those who were working long hours on essential work and those who were medically unfit. The final category which could be exempted were those for whom, if they performed these duties would incur 'exceptional hardship'. At first the Business Order, which was amended later in 1941, presented no difficulties for conscientious objectors because employers could engage voluntary work for fire-watching under the Order and objectors could therefore continue to fire-watch voluntarily. But the Compulsory Enrolment Order involved the registration of all male subjects between the ages of 16 and 60 in order that they could be called upon to undertake compulsory fire-prevention duties under local authorities.

The position of some conscientious objectors on the matter of compulsory fire-watching was much less easy for the public to understand than that on, for instance, industrial direction. Most people could accept that a declared and accepted conscientious objector should not be required to undertake munitions work. But that a conscientious objection could be entertained to fire-watching appeared unreasonable and irresponsible. Surely it was first and foremost work of a humanitarian nature. However, in almost all cases, it was not an objection to fire-watching as such, but an objection to the compulsion to fire-watch which objectors entertained. This was a distinction which the State could hardly recognise and no provision for a conscientious objection to compulsory fire-watching was made in the Orders. The CBCO met later in January and discussed the situation. The Board knew for sure that there would be objections from some conscientious objectors on this matter. It decided that on no account should these objections be encouraged but that support should be given to those who found that they had a conscientious objection to compulsory fire-watching. Stuart Morris was asked to approach the Minister of Home Security on the matter, and the Exemptions Group were asked to press for a statutory right to object conscientiously to compulsory fire-watching.(52) The Minister was non-committal. He wrote to the CBCO saying that there was no reason why objectors should not state that they had a conscientious objection to compulsory fire-watching when applying on the grounds of exceptional hardship to the Exemptions Tribunals. But he felt that he could not instruct the Tribunals on 'how such applications should be treated'.(53) Although a very small number of Tribunals allowed a conscientious objection to come under the heading of 'exceptional hardship', most did not, and prosecutions of objectors for either refusing to register or for refusing to carry out fire-prevention duties once directed became increasingly frequent.

As usual deputations were arranged to see officials of the relevant Ministry, in this case, the Ministry of Home Security. The first of these occurred at the end of 1941 when three members of the Exemptions Group, T. Edmund Harvey MP, Campbell Stephen MP, and Rev. J. Barr MP met Ellen Wilkinson and Osbert Peake, Joint Parliamentary Secretaries of the Ministry. Harvey began by saying that Tribunals were not sure whether they could consider conscientious objection as exceptional hardship and that he felt that this ought to be made clear, one way or the other, by the Ministry. He also complained that the penalties awarded by the Courts for refusal to fire-watch were widely divergent; some offenders were nominally fined, others were imprisoned for the term of three months. Guidance on penalties, he thought, should be given by the Ministry. He commented that had the matter of compulsory fire-watching been the subject of an Act of Parliament rather than an Order under the Defence Regulations, a number of amendments would have been passed. He particularly wanted the Home Office to discontinue repeated prosecutions of offenders and said that many conscientious objectors were perfectly prepared to carry out fire-prevention duties voluntarily. Miss Wilkinson replied that voluntary fire-watching was not enough. The minutes of the meeting recorded her words: 'it was necessary to have an organised service to prevent the country being burned down. It was the lack of such a service which had led to the terrible fires in London, Plymouth and Bristol.' It was impossible, she said, to make refusal to fire-watch a non-continuing offence since there were many people, not conscientious objectors, who would be tempted in that case to refuse fire-watching duties. Osbert Peake must have spoke on behalf of many people when he said that he just could not understand why conscientious objectors refused to fire-watch under compulsion when they were prepared to fire-watch voluntarily. In any case he promised to place before the Minister the deputation's representations.(54)

A month later Miss Wilkinson and two colleagues at the Ministry of Labour received another deputation, this time from officers of the CBCO. Joe Brayshaw, on behalf of the Board, said that the Minister had commented that there was no reason why conscientious objectors should not submit their cases to the Tribunals although he was in no position to advise them on the matter. Miss Wilkinson said that the Department had now been advised by its legal advisors that the Tribunals had no power to consider conscientious objectors' claims in connection with applications for exemption for reasons of hardship. The officers of the CBCO pressed that there should be official clarification on this matter and then every interested party would know where they were on this matter.(55)

Indeed it was this confusion on points of law, or the lack of them, which was the fundamental difficulty in dealing with conscientious objections to compulsory fire-watching. Two days after the deputation from the CBCO, a civil servant in the Ministry of Labour commented in a memorandum to a colleague, 'the absence at present of a clear statement of the law on this subject is certainly embarrassing.'(56) The maze of legislation and Orders on the subject of fire prevention continued to confuse almost everyone until the end of the war. Conscientious objection to compulsory fire-watching was never admitted and a considerable number of objectors were prosecuted more than once for failing to undertake fire watching duties. Denis Hayes relates the more notorious of these cases; the nine prosecutions of the Jehovah's Witness, George Elphick, for his failure to fire-watch in Lewes were given much publicity at the time.(57) In total 475 conscientious objectors were prosecuted at least once for offences connected with fire-watching during the war.(58)

The lack of success of the CBCO and the Exemptions Group in obtaining a statutory right of conscientious objection to compulsory fire-watching or even administrative concessions such as those allowed in the case of 'cat and mouse' or industrial direction can be ascribed to two interrelated factors. Firstly, the Government and public opinion could not be convinced that it was possible that a conscientious objection could be entertained to compulsory fire-watching. They reasoned that, as it was obvious that conscientious objectors had no objection to the work itself because they had been seen to perform it willingly on a voluntary basis, conscientious objectors were objecting merely to compulsion. One might object to compulsion, but could one conscientiously object to compulsion? The objectors concerned would have argued that the compulsion resulted from the prosecution of the war and was made necessary for the successful prosecution of the war; as they conscientiously objected to the war, they must conscientiously object to compulsion for war purposes. All the old objections could be made to this argument but perhaps this time more appositely: objectors were compulsorily taxed, and much of the money raised went towards war purposes, yet they did not conscientiously object to that. There had been no conscientious objections to compulsory 'black-outs'.

The second factor which may account for the failure of the CBCO to make any real headway on the matter of fire-prevention duties was that many conscientious objectors, including some who worked for the

Board, were not entirely convinced that an objection to compulsory fire-watching was conscientious objection. The CBCO always respected the consciences of those who refused and undoubtedly did as much as was possible to support, help and advise them. But the case was not anywhere near as clear cut and understandable as that for the prevention of 'cat and mouse' or conscription of conscientious objectors to war work, and officers of the CBCO realised this. Hence their decision that this was a question for the individual conscience alone; the CBCO itself took no stance on the matter except to plead the case that the laws and Orders relating to fire-prevention duties be clarified. Probably no other issue illustrated more clearly the individual nature of conscience and the widely divergent stands which that individual conscience led objectors to take.

One other new development occurred during the war which concerned conscientious objectors, and therefore the CBCO, and that was the conscription of women introduced at the end of 1941. Although in the First World War women had shown that they could do more for the war effort than merely look after the homes and children of the men in the Services by performing work hitherto the exclusive province of men, military conscription for women had been unthinkable. Judging by the outcry in Parliament when the proposals of the Government were announced, many still felt it unthinkable.(59) However, in practice the measure resembled industrial conscription rather than military conscription. There were Women's Services but women were free to choose between them, Civil Defence Services or work in industry. Additionally, no woman would have to use a 'lethal weapon' unless she gave her written permission. Conscription was limited in any case to single women between the ages of 19 and 31, and in fact only those women born between 1918 and 1923 were called up. The question, then, for the CBCO was how the administration of the women's 'conscience clause' would be handled. The 'conscience clause' was exactly similar to that in the earlier National Service Act except that there was provision for at least one woman to sit on a Tribunal which was hearing the case of a woman conscientious objector. If a woman claimed to be a conscientious objector on registration, she would often be offered civilian work, and if she had no objections she would proceed to that work without having a Tribunal hearing. However, the CBCO were anxious to point out to women that if they wanted a Tribunal hearing even though they were prepared to undertake the work offered, they were entitled to it and should insist upon it.(60) Thus only those who either objected to the work offered or were determined to have their conscientious objection formally acknowledged appeared before the Tribunals. The wish for a formal acknowledgment was not sheer pedantry by women; it might well affect their position in relation to Civil Defence, fire-watching and industrial duties in the future. Nevertheless, many women objectors chose not to appear before the Local Tribunals.

Once the CBCO established that the conscription of women, as far as the 'conscience clause' and the Tribunal hearings were concerned, was to be administered in the same manner as that of men, they treated the individual woman's case which was apparently unfair in exactly the same manner as that of a man. Just over 1,000 women appeared before Local Tribunals during the war and just under half that number appealed against their decisions,(61) a proportionally greater number than the men who appealed. A smaller number of women served prison sentences. A large number of appeals may be accounted for by the fact that many of the women who took the trouble to appear before the Tribunals were unconditionalists, determined to make a formal stand.

The cases of industrial direction, compulsory fire-prevention duties and the conscription of women, in addition to those of 'cat and mouse' and compulsory Civil Defence duties related in other chapters, show how the CBCO and the Exemptions Group helped to frame, amend and secure fair legislation in relation to conscientious objectors. They also managed on occasion to persuade Ministers to make administrative concessions which greatly alleviated some problems of conscientious objectors. Although most of the help given to individuals was at local level, the Board and the Exemptions Group were able, between them, to give immense publicity to any case which they felt warranted it. However, the CBCO was not an entirely passive organisation, giving help and advice where it had become necessary; from it sprang ideas which the officers did not hesitate to present to the Government either through the Ministries or in Parliament.

An example of this was the pressure exercised to persuade Allied Governments to adopt Britain's liberal attitude to conscientious objectors and to introduce a 'conscience clause' into their legislation for conscription. The first mention of this idea in the Board Minutes occurred in August 1941 when it was reported that letters had been sent to all the Governments of the Allied countries urging them to adopt a 'conscience clause'.(62) Towards the end of September replies had arrived from some countries. The Norwegian Government said that it already had an exemption clause included in its legislation and the Netherlands maintained that conscientious objectors had been exempted there since 1923. The Polish Government in London simply acknowledged the letter and somebody working for the Free France movement wrote to say that Free France did not intend to provide for conscientious objectors. It was agreed to write once more to those Governments which had not replied.(63) Presumably they never replied for there is no further mention of the subject in Board minutes.

Another example of this type of activity was when Fenner Brockway, chairman of the CBCO, suggested that objectors with civilian employment should give away part of their salary so that they would earn no more than if they were soldiers. He wrote to the Ministry of Labour in July 1940 expressing concern about the widespread victimisation of employees of local authorities.(64) One of the main reasons he felt that the public resented conscientious objectors was that they were likely to earn more in civilian employment than if they were in the Armed Forces. He therefore wrote that the CBCO had recommended the proposal that

> they [conscientious objectors] should voluntarily agree to contribute to some relief of the victims of war or some social purpose of their own choice the difference between the living standards of their fellow workers who have joined the Forces and their own wages or salaries.

He stressed that the system would have to be voluntary because some absolutists might object.(65) Later in the month Fenner Brockway and other officers of the CBCO went to see the Minister who 'greatly appreciated' the offer and approved of it but felt that if the idea were to be adopted, it should be done compulsorily.(66) Mention of compulsion to a group of conscientious objectors, however, was like waving a red flag at a bull. The Board wrote a long letter to the Minister citing the multifarious reasons why the scheme should never be compulsory. Firstly, the Board could hardly endorse the principle of compulsion when most of its time was spent fighting that very concept. Secondly, the idea of compulsion destroyed the whole idea of the scheme: it was designed to show that the conscientious objector would willingly, of his own accord, make a sacrifice to show that he did not wish to benefit financially from the privilege of being exempted from the Armed Forces. Thirdly, it was obvious that some objectors would be able to pay more than others and a scheme of compulsion might result in extreme hardship for some other objectors. In any case, what of those who were in civilian occupations because they were medically unfit or were in reserved occupations? Were they to be included too in this compulsory scheme? Fourthly, a scheme of compulsion would involve acceptance of the principle 'of recognising that the Government has a right to discriminate against them (conscientious objectors) by reason of their holding particular views.' Lastly, it would mean a financial inquiry into many people's lives which, apart from any moral objections, would involve considerable administrative machinery.(67)

Despite these objections Ernest Bevin seriously considered the possibility of a compulsory scheme. He even mentioned it to the War Cabinet telling Ministers that the dismissals were worrying those, and not necessarily pacifists, 'who see in the discrimination against Conscientious Objectors purely on account of their convictions a denial of the very ideals for which we are supposed to be fighting.' (68) But when he presented the scheme to the TUC it was met with blank disapproval. It conflicted with every trade-union ideal of the 'rate for the job'. So in February 1941 he made the following statement in the House of Commons in answer to a question from T. Edmund Harvey:(69)

> I have given careful consideration to this proposal which would require legislation before it could be put into effect, and have discussed it with the Joint Consultative Committee representing the TUC and the British Employers' Confederation. As a result, I am satisfied that, whatever may be the merits of the proposal otherwise, it would arouse acute controversy and would require disproportionately elaborate and expensive administrative machinery for its effective operation. In the circumstances, I have decided not to proceed in the matter.

Why the CBCO could not organise its own voluntary scheme is not clear. Assuming that the idea was presented to the Minister in all sincerity and that it was not merely a piece of astute public relations work, it may have been that the CBCO felt that it could not organise its own scheme for much the same reasons as Bevin gave. Collecting and administering payments might have taken too much of the limited time officers had to give to the Board. An additional factor was that, as 1941 drew on, public resentment towards objectors lessened and the need for such a scheme had disappeared. But the gesture, which was given much publicity, had been made.

There is no doubt that the CBCO provided an invaluable service for conscientious objectors during the Second World War. Most of the money it raised came from private donations and from the sale of its many publications. There was no payment required to elicit the help of the CBCO. Indeed the Board was sometimes in financial difficulties; in 1943 a Special Appeals Week was arranged for September because money was so short.(70) Despite these problems the Board continued with its undoubtedly well-intentioned and devoted work through the war and afterwards, for conscription did not end with the war. Fenner Brockway comments, 'without any qualification, I can say that in forty years' experience of innumerable councils and committees I have never known such a harmonious body as the Central Board for Conscientious Objectors.'(71) Much other work was done by societies affiliated to the CBCO but the advantage of the Board was that it could co-ordinate all action in relation to conscientious objectors. It also stuck admirably to its policy of refusing to make propaganda of pacifist principles despite accusations to the contrary. In 1941 the Board adopted unanimously the statement:(72)

> In view of mis-statements in the press the CBCO reaffirms its policy of support for all who, from their own convictions, take the stand of Conscientious Objectors. The Central Board does not advise anyone to take any attitude to military service but that which his own conscience dictates.

And this policy was adhered to, for a year later during a discussion on public relations, the following resolution was adopted: 'that anyone acting as a representative of the Board should not attempt to direct an individual as to how his conscience should operate.'(73) The CBCO was a service organisation, never a militant propaganda machine, and it remained so throughout the war.

CONCLUSION

In framing legislation for the provision for conscientious objection to military service, the lawmakers of the Second World War had two major advantages over those in the First. In 1916 the question was whether or not to introduce legislation; in 1939 the question was how to frame it. The difference in emphasis was crucial. In the furore of the achievement of including a 'conscience clause' at all in the Military Training Act in 1916, insufficient attention was given to the problem of how such a clause should be administered. Many of the abuses and injustices of the First World War were directly attributable to this circumstance. In 1939, when the concept of a 'conscience clause' in legislation introducing conscription was widely accepted, proper attention could be devoted to the detailed problems: where should applicants be 'tested', who should 'test' them, what conditions of exemption should be allowed, to what use could the labour of genuine applicants be put, what penalties should be imposed on unsuccessful applicants who refused to 'join up'? The second major advantage was that, in answering these questions, the experience of the First World War could be drawn upon. No such facility existed in 1916.

By far the most serious mistake of the First World War was to involve the Armed Forces in the administration of the 'conscience clause'. In 1939 this reponsibility was given from the first solely to the Ministry of Labour and National Service under the aegis of Ernest Brown and, from 1940 to the end of the war, Ernest Bevin. Thus the handling of conscientious objectors was a civilian affair. Even an unsuccessful applicant, if he refused to undertake the medical examination, need never have had any contact with the fighting forces. Much of the tension of the First World War was avoided, therefore, in the Second. The worst results of the uneasy marriage of military men and conscientious objectors were evident in the Tribunals which in 1916-19 were administered by the War Office. Meticulous attention to detail was paid in the clauses of the 1939 Military Training Bill which set up the Tribunals; guidance notes for chairmen and members were issued. Legislation provided a tight framework so that members could be free to make their decisions without being too bound by formal procedures.

The decisions that Tribunals could make were based on the format of the First World War. It was not simply a question of persuading members of the genuineness of an application; the question of degree of objection was thought by the lawmakers to be vitally important. Members had to decide exactly to what an applicant objected. Most applicants were exempted on condition that they undertook civilian work under civilian control but it is highly likely that, once they had accepted an applicant's objection as conscientious, members ordered this exemption because it was the area in which they had most power to direct objectors into activities which would assist the war effort rather than because they had judged exactly the degree of objection.

The effective choice, given that the applicant was thought to be genuine, and unless unconditional exemption was given which was rare, was between conditional exemption and entry into the Armed Forces as a non-combatant. But in wartime the value of soldiers, who were in no circumstances prepared to use lethal weapons, was limited. There were, of course, many non-combatant tasks to be undertaken in the Armed Forces, but they were so inextricably interwoven with the ultimate purpose of disposing of the enemy that it would have been an exceedingly difficult task to find enough useful non-combatant work for a significant number of men. The Tribunals therefore preferred to give conditional exemption.

It was clear from the Military Training Bill that the Government's aim was to recruit Tribunal chairmen and members who would give applicants as impartial a hearing as was possible. This arose from criticisms of hearings in the First World War where accusations that members were actively attempting to recruit for the Armed Forces were not unfounded. Not surprisingly, the legal profession was thought to be the most productive source of impartial professional men. Judges, barristers and solicitors made up all the chairmen of the Tribunals and many of the members as well. Other members were highly respected men, too. Representatives of the TUC came from the higher echelons of the organisation.

Still the question remained: was it possible to 'test' conscience at all? It has been argued that Tribunals were only able to judge the sincerity of an objection, and that they attempted to do so by a series of probing questions which they hoped would reveal the depths and degree of an applicant's objections. If the applicant appeared sincere, then it had to be assumed that the objection was conscientious. In practice this approach worked fairly well. There were obvious pitfalls, however. Personality would, perhaps, loom too large in the assessment of the applicant's convictions. Some genuine applicants may not have had a disposition which made it easy for them to appear sincere. Inarticulate applicants would be at a disadvantage. Some ideas aired may have been so difficult for the Tribunals to understand, let alone sympathise with, that they would have been forced to the conclusion that the applicant could not possibly have been sincere. On the other hand, sincerity is not difficult to feign; 'con men' would be out of business if this were not true.

The correctness of the Tribunal decisions made on this basis can at least be partly evaluated by considering firstly, the number of appeals against the decisions, and secondly, and more particularly, the number of successful appeals. The first shows how many applicants were dissatisfied with their decisions, the second, how many decisions Appellate Tribunals also felt were wrong. Of all the applicants, about 30 per cent appealed against their decision whether their application had failed altogether or whether they had been placed in a category of exemption to which they objected. Roughly 50 per cent of these decisions were varied by the Appellate Tribunals. Where Local Tribunals had decided that the applicant was genuine, the Appellate Tribunal disagreed in only 1.4 per cent of cases. But where the Local Tribunals had decided that the applicant was not genuine, they disagreed in nearly 50 per cent of cases.(1)

This is merely a guide to the success of Local Tribunals in getting their decisions right. Appellate Tribunals were far from infallible either and bore little relation to appeal courts in the legal system. Appeal courts have an opportunity to read the transcripts of the evidence and summing up of the cases in the first hearing. The Appellate Tribunals were more akin to a rehearing since they had little more information than simply the decision that the Local Tribunal had made. Nevertheless it is apparent from the figures that those who appealed had a 50 per cent chance of having their condition varied, and if their application had failed, a 50 per cent chance of being awarded a condition of exemption in the Appellate Tribunal. Similarly, in law courts very few barristers or solicitors would recommend appealing against a decision in the lower court unless there was a 50 per cent chance of a different decision in the appeal court. Yet the reputation of lower courts as a whole is not tarnished by the success of appeals against its decisions. The very existence of appeal courts acknowledges that lower courts will make mistakes. All that is expected is that the mistakes will be the exception rather than the rule, and this was largely true of the Tribunals of the Second World War.

Predictably, most applicants gave their reasons for objecting as religious although as Hayes pointed out, this was often because, if an applicant had a religion, he would cite it on his application form even though the basis of the objection was not specifically religious.(2) However, there was inherent resistance in some faiths to military service, for instance, in that of the Quakers and the Jehovah's Witnesses. And there were also those who, despite their church leaders' support of the fighting, interpreted the doctrines of their faith in a manner which, they felt, precluded their fighting. There is no doubt that applicants with religious objections found it easier to convince Tribunals of their sincerity than those whose objections sprang from moral, humanitarian or political beliefs.

While it was thought that mainly 'leftie' intellectuals, students and teachers constituted the main body of conscientious objectors, it was quite simple to offer an explanation for it. They had access to books, to political treatises, they perhaps considered at length their relationship with the state. They were intelligent. But how can the preponderance of 'white-collar workers', non-intellectuals, lower middle class, semi-professional and self-employed men and women be explained? This group probably accounted for a majority of the religious objectors; their sociological 'band' provides the corpus of most church congregations even now and probably more so in 1939 when congregations were larger. The self-employed were likely to have an independence of spirit not easily cajoled into something to which they objected. However, these are only possibilities and there is room here for a more penetrating sociological study of the phenomenon.

The Tribunals of the Second World War were a distinct improvement over those of the First, and this was largely due to the experience of the First World War. Enormous care was taken in framing legislation for the setting up of the Tribunals; they were given a judicial air, an entirely non-military personnel and formal procedures which went some way towards ensuring that the applicant was given a fair hearing. However, there was ample opportunity for abuse of the system. Chairmen, always county-court judges, on occasion became emotionally involved, insulting and bullying applicants, both before and after their decisions had been made. Although applicants were allowed solicitors or counsel to represent them, few did so, probably because of the expense involved. There was, therefore, little check on the manner in which Tribunals questioned applicants and indeed on the questions and remarks themselves. Initial feelings of hostility were probably greater in a Tribunal than in a criminal court, where at least the motives, although deplored, could be understood. But outright hostility was confined to a few Chairmen. In general, Tribunals did their best to give the applicants every chance to convince them of their sincerity. Within the limitations of the system the Tribunals performed efficiently and equitably.

For the applicants, their Tribunal hearing was just the start of their experience of the Second World War as conscientious objectors. Most of them were exempted on condition that they undertook civilian work under civilian control, work of 'national importance', a proviso which the Government, the Tribunals and the objectors found difficult to interpret and obey. Work in agriculture had seemed at the beginning of the war to be an obvious choice but it was fraught with difficulties, due partly to an uneven distribution of available jobs and partly to prejudice against employing conscientious objectors. In any case the objectors were very often ill-suited to the work to which they had been directed. This applied to other work which was described as work of 'national importance'. And there was the danger of trade-union hostility if objectors started taking jobs normally the province of their members.

In many cases it would probably have been of more use to the community to have left objectors in their own occupations for which they were trained and in which they had experience even though the job might not have been of primary national importance. However, there was a punitive aspect to the problem. If conditionally exempted men were allowed to continue in their own jobs, they might just as well have been unconditionally exempted. It hardly seemed fair that while the rest of an objector's age group was being enlisted for the Armed Forces, and therefore being uprooted from home and family, the objector should be allowed to continue in his own home and work environment. A partial solution to this problem was found by a few Tribunals later in the war when they specified spare-time voluntary work in addition to continuing with the existing occupation as a condition of exemption. This involved a degree of 'hardship' but also made additional use of the objector's labour for the good of the community.

In practice hardships were, in any case, involved. Even those who had been instructed to carry on in their own jobs sometimes found that their employers no longer desired their services in the light of their convictions. New jobs, especially where they had no experience or skill, were hard to find in the face of prejudice. Unemployment among objectors at the beginning of the war was a problem. However, in both the public and private sectors, much of the antagonism towards conscientious objectors disappeared by the end of 1941 when their labour was more in demand and when many of them had had a chance to show that they were not all 'shirkers' and 'funks'.

Imaginative uses of objectors' labour were occasionally found, however. Dr Kenneth Mellanby's work (3) using conscientious objectors as guinea-pigs for a cure for scabies is described in his book 'Human Guinea-Pigs'.(4)

Humanitarian work, which objectors were particularly keen to perform, was later prescribed by the Tribunals. Objectors in the First World War had set a precedent for this; some 6,000 out of a total of 16,000 performed civilian work and 1,000 of these worked for the Friends' Ambulance Unit.(5) Hayes believed that in the First World War the 'service of society had been of secondary importance'. 'In the name of their stand they virtually outlawed themselves for the duration of the war.' This fight to beat the 'evil system' enabled conscientious objectors of the Second World War, continued Hayes, to perform the service to the community which circumstances had denied those of the First.(6) Circumstances were, of course, different, though whether the pattern of 'witness' in the First World War and 'service' in the Second was consciously followed by most objectors is doubtful. In any case the work successfully undertaken by conscientious objectors in the First World War was repeated and expanded.

The debates in local authority chambers on whether or not to retain objectors in their employment, and on what terms, are illuminating. Due in most part to pressure from the public, they took place in 1940 and 1941 when hostility towards objectors was strongest. They showed how ordinary people felt about conscientious objectors and how many of them were torn between their revulsion against men and women who were refusing to fight in this most urgent of wars, and their respect for central Government which continually urged employers not to discriminate against objectors. In addition, it was often pointed out in debates that to persecute this minority was not unlike some of the activities to which Hitler was prone and against which Britain was fighting.

Legally there was no compulsion upon local authorities either to retain or engage conscientious objectors. The question in the debates, then, was a moral one: was it right to dismiss objectors and not to employ them? In a few cases blind prejudice decided the issue. But in most a genuine dilemma existed. The authorities had a public duty to perform and were accountable to the public. If that public demanded that its rates should not be spent on employing conscientious objectors, then authorities were forced at least to discuss it. Many of them wanted to be fair; to dismiss objectors seemed tantamount to Hitlerism, to retain them seemed unfair to those employees who had been called up into the Armed Forces.

The intention of the law had been that the Tribunals should

decide the fate of each conscientious objector. Local authorities felt differently, however. A large proportion took action of one sort or another against objectors. But, again, by 1942, authorities had started to employ objectors for much the same reasons as the farmers. Their services were in demand, especially in Civil Defence; the disapproving voice of central Government was at last being listened to and the public attitude towards objectors had changed. Even objector teachers were finding it easier to retain and gain employment at that stage in the war.

Conscientious objectors in civilian occupations, then, faced enormous prejudice but, if they overcame it, could perform a useful and fulfilling job of work. For those objectors who were non-combatants the situation was rather reversed. Surprisingly little persecution of objectors occurred in the Armed Forces. The separation of the military from the politics of conscientious objectors must have helped in this. In addition the military was less susceptible to fashions of public opinion and, because it was directly involved in combating the German threat, did not feel so much of the frustration that civilians suffered, a frustration which the presence of conscientious objectors appeared to aggravate.

On the other hand, the value of the work of the non-combatants was always in question, especially by the objectors themselves. Though numerous tasks were found for them on paper, in practice they were underused and ill-suited to the tasks they were given. And although their objection was merely to the act of killing, being a part of an organisation designed for just that purpose must have made for a hollow and uncomfortable existence. Possibly the Armed Forces was not the place for them at all.

Those who declared a conscientious objection to military service once they were in the Armed Forces, as in the First World War, caused problems. 'Cat and mouse' seemed likely to dominate the history of conscientious objection at one time, as it had done in 1916-19. But because so many people, not just objectors and their sympathisers, but also objective MPs, churchmen and Government Ministers remembered so well the injustices of the Great War, steps were taken to avoid its recurrence. Provision in the original Act was found to have loopholes but by 1943 most of these had been closed, not by more legislation but by strong leads from central Government, especially the War Office. Another lesson had been learnt from the First World War.

Hayes estimated that three out of every ten objectors in the First World War spent some time in civil or military prisons.(7) In the Second World War the figure was less than one in ten. That offences, apparently committed on the grounds of conscience and where the offender was quite obviously of no danger to the public, should have been punishable by prison sentences seems, in retrospect, ill-conceived. But a prison sentence at that time was still very much more of a punishment than a rehabilitation, and was also thought to act as a deterrent, not so much against objectors but against those unscrupulous characters who might have taken advantage of more lenient punishment for offences supposedly committed on conscientious grounds. Objectors once in prison were rarely ill-treated although conditions in HM Prisons in wartime were exceptionally unpleasant. Difficulty in finding acceptance and employment after their prison sentence was served was the effective punishment.

Of the organisations concerned with conscientious objectors the one most remembered is the PPU. A quasi-political organisation which actively campaigned both before and during the war for an end forever to conscription and to war, the PPU gained a notoriety never enjoyed by the CBCO, the main organisation representing objectors. The CBCO was a creation of the Second World War; any organisation connected with or interested in the welfare of conscientious objectors could affiliate to it and, in this way, action on behalf of all objectors could be co-ordinated centrally. Many highly respected persons had been pacifists during the last war and between the wars. Bertrand Russell, George Lansbury and indeed the Minister of Home Security in 1940, Herbert Morrison, had all been active pacifists. This respectability helped the officers of the CBCO to win the confidence of civil servants and Government Ministers, aided of course by the elder statesman of pacifism, Fenner Brockway. Throughout the war, increasingly good relations betwen the CBCO, the Government and an Exemptions Group of MPs sympathetic to objectors, helped the administration of the 'conscience clause' immeasurably.(8)

At the end of the war the release of objectors both from the Armed Forces and from Tribunal-directed civilian work was planned very much on the lines of the demobilisation of the fighting forces. The new Labour Government in which George Isaacs, a trade-union official, was appointed Minister of Labour (Bevin moved to the Foreign Office), introduced the National Service (Release of Conscientious Objectors) Bill in October 1945. The Bill had a difficult passage, especially in the House of Lords, but in 1946, when it was enacted, its main provisions had survived. The Minister of Labour had power to release objectors from their Tribunal conditions at roughly the same time as officers and men in the Armed Forces in the same release group. The group was calculated on age and length of service. Objectors who had served terms in prison and those who had been exempted from military service by the Advisory Tribunal were released similarly. Peacetime conscription continued until 1959 and the experience of objectors in the intervening years is described in Hayes up to 1948,(9) and in a small pamphlet by Constance Braithwaite, 'A Short History of the Legal Position of Conscientious Objectors to Military Service and Other War Time Compulsions (1964)'.(10)

The main body of conscientious objectors was, therefore, released in 1946 and 1947. How should the Government's introduction and administration of the 'conscience clause', the experience of the conscientious objectors and the reaction of the Armed Forces and the public to them be assessed? Much was owed to the careful drafting of the original Bill. This set a tone and framework from which fair administration and further legislation could emanate. The Tribunals, their task an impossibly difficult one and, in many ways, an unsatisfactory forum for 'testing' conscience, coped successfully within their limitations. The Armed Forces were removed entirely from the political side of the issue and were probably glad of it. Their relations with objectors in the Armed Forces were generally good. Despite central Government pressure, the public remained hostile to objectors. A feeling of antagonism would have been more than bearable; after all, the Government could hardly expect, on the one hand, to whip up in His Majesty's subjects a war fervour, loyalty to one's country and the will to defeat the enemy and, on the other, to ensure that those same subjects should accept into society without a murmur those who appeared to take an opposite view. However, the public's power to make it very difficult for objectors to retain and obtain employment was in every way counter-productive.

For the smooth administration of the 'conscience clause' the co-operation of the conscientious objectors themselves was essential. Had they, for instance, refused to attend their Tribunal hearings in large numbers, the system could well have collapsed. However, most were content to accept the dictates of the Tribunals' decisions and even to accept the increasing demands made on their freedom by the Government throughout the war. A responsible and, importantly, an a-political organisation for conscientious objectors, the CBCO, did much to ease relations between the minority and the Government. Ernest Bevin, who for the most part of the war had responsibility for objectors, proved to be a most suitable man for the job. Pragmatic, business-like and a superb administrator, his influence in the Ministry of Labour defused potentially explosive situations before they happened.(11)

In the Second World War the 59,000 conscientious objectors represented roughly 1.2 per cent of the 5 million 'called-up'. In the First World War it is estimated that there were 16,000 objectors,(12) and this represents about 0.125 per cent of the 6 million enlisted, although that figure includes volunteers before conscription in 1916. The proportion would be slightly higher if the figures included only those conscripted. There was, then, a marginal increase in the number of conscientious objectors in the Second World War. This small rise may be explained by the measure of public acceptance of pacifism between the wars and by the fact that in 1916 conscientious objection was a relatively new concept in British political life while, by 1939, it was a part of it. Pacifists between the wars would, therefore, expect to become conscientious objectors in any war.

While Great Britain is no longer obliged to have a policy for conscientious objectors, since conscription is no longer in force, for many countries the issue is still very much alive. 'Draft-dodgers' in the USA, although not strictly conscientious objectors, won world-wide attention during the war in Vietnam in the 1960s. Most European countries made no provision at all for a conscientious objection to military service during the war. Openly declared objectors in Austria, Belgium, Bulgaria, Czechoslovakia, Finland, France, Germany, Hungary, Italy, Norway, Poland, Portugal, Spain, Russia and Yugoslavia risked imprisonment and execution. Denmark managed to operate an alternative service law even during German occupation, the Netherlands similarly had some provision, Sweden made provision and the USSR, interestingly, had had some provision until 1939 when it was cancelled.(13)

At the present time most Western European countries which still conscript have made some provision for conscientious objectors, although in comparison with the standard set by Great Britain in the Second World War, of a limited nature.(14) Eastern European countries are less generous and work is now being done by, among others, the Society of Friends and Amnesty International to relieve the suffering of those conscientious objectors in prison. The achievement of Great Britain in making such comprehensive provision for conscientious objectors during two world wars showed a relaxed and secure society in which the state could acknowledge that some of its individuals could not, in all conscience, obey all of its laws.

NOTES

INTRODUCTION

1 Denis Hayes, 'Challenge of Conscience', 1949.
2 John Hughes, The Legal Implications of Conscientious Objection, MLl thesis, University of Manchester, 1971.
3 Clifford Simmons, 'The Objectors', 1965.
4 Peter Brock, 'Twentieth Century Pacifism', 1970 and Martin Ceadel, 'Pacifism in Britain 1914-45', 1980.
5 Thomas Hobbes, 'The Leviathan', 1651.
6 John Locke had, in the seventeenth century, set a tradition for the nineteenth-century philosophers, J.S. Mill and Jeremy Bentham, of the need for state tolerance of the individual. Adam Smith, although concerned with the economic welfare of the nation, continued in the tradition of the need for a minimum of state interference.
7 For instance, see Aldous Huxley, 'Ends and Means', 1931, Bertrand Russell, 'Justice in Wartime', 1916 and George Bernard Shaw, 'What I Really Wrote About the War', 1931.
8 See Ceadel, op.cit., for an account of the great pacifist writers in this period.
9 Bertrand Russell, Bernard Shaw, A.A. Milne and Cyril Joad were among those who publicly recanted their pacifism just before or during the war.
10 See Brock, op.cit., Chapter 6, for an introduction to the development of pacifism after the Second World War. Also see Michael Howard, 'War and the Liberal Conscience', 1977, for a history of liberal attitudes to war.
11 G.C. Field, 'Pacifism and Conscientious Objection', 1945.
12 Ceadel, op.cit., has made a further distinction between pacifism and pacificism, the first being a moral objection to all war and the second an objection to war with a reluctant acceptance that wars must sometimes be fought. The Second World War brought out countless latent 'pacificists' who accepted, in the end, that Hitler's war was a war worth fighting. In addition to those in note (9), Maude Royden and Leslie Weatherhead, both preachers, left the PPU at the beginning of the war and the watershed, when 'real war' came in 1940, produced PPU resignations from Philip

Mumford, a former professional soldier, and the novelists, Margaret Storm Jameson, Rose Macauley and J.D. Beresford among others. See Ceadel, op.cit., pp.221-4 for details.

13 A. Campbell Garnett, Conscience and Conscientiousness, in Joel Feinberg (ed.), 'Moral Concepts', 1969.
14 Peter Singer, 'Democracy and Disobedience', 1973, pp.92-5.
15 Ibid.
16 See particularly Cecil Cadoux, 'Christian Pacifism Reexamined', 1940, G.H.C. MacGregor, 'The New Testament Basis of Pacifism', 1936, Charles Raven, 'War and the Christian', 1938, Leyton Richards, 'The Christian's Alternative to War: An Examination of Christian Pacifism', 1929, F. Stratmann, 'The Church and War: A Catholic Study', 1928 and W.M. Watt, 'Can Christians be Pacifists?', 1937.

CHAPTER 1: PROSPECTS OF WAR: INTER-WAR PACIFISM AND THE INTRODUCTION OF CONSCRIPTION

1 Its membership reached a peak in 1931 when 406,868 subscriptions were collected, Martin Ceadel, 'Pacifism in Britain', p.317.
2 Michael Howard, 'War and the Liberal Conscience', p.88, C.L. Mowat, 'Britain Between the Wars 1918-40', 1955, pp.541-2 and Adelaide Livingstone, 'The Peace Ballot: The Official History', 1935, p.34.
3 Ceadel, op.cit., pp.177-8. For a contemporary account of the beginnings of the Peace Pledge Union, and a full version of the letter, see Sybil Morrison's 'I Renounce War: The Story of the Peace Pledge Union', 1962, pp.99-100.
4 Howard, op.cit., p.90.
5 Hector McNeil, Foreign Policy Between the Wars, in 'The British Labour Party', vol.2, (ed.) Herbert Tracy, 1948, p.267.
6 A.A. Milne, 'Peace with Honour', 1934 and 'War with Honour', 1940.
7 Ceadel, op.cit., pp.296-7.
8 See Herbert Morrison, 'An Autobiography', 1960. John Rae, in 'Conscience and Politics', gives other examples of the fate of some concerned with conscientious objection in the Great War from p.240 onwards.
9 F.W. Pethick Lawrence, 'Fate Has Been Kind', 1942, p.118.
10 Ceadel, op.cit., p.180.
11 Ibid., see pp.222-41 for a catalogue of other well-known activists and sympathisers in the PPU at the beginning of its life.
12 Ibid., p.263.
13 'Hansard', 310, 1992, 1 April 1936.
14 A.J.P. Taylor, 'English History 1914-45', p.480; Mowat, op.cit., p.570; Denis Hayes, 'Conscription Conflict', 1949, p.370. Baldwin was accused of deceiving the electorate for he had promised during the 1936 general election that there would be 'no great armaments'.
15 Mowat, op.cit., p.632.
16 'Daily Mail', 31 May and 2 June 1938 and 'Hansard', 336, 1765-78, 30 May 1938.
17 'Hansard', 336, 2016, 1 June 1938.

18 Ibid., 345, 290-2, 14 March 1939.
19 Hayes, op.cit., pp.376-8.
20 Ibid., p.379.
21 Ibid., pp.373-4.
22 'Daily Mail', 25 July 1945.
23 Hayes, op.cit., p.380.
24 PRO, CAB 23, 14(38)4, 16 March 1938.
25 Ibid., 15(39)5, 29 March 1939.
26 Ibid., 22(39)3, 24 April 1939.
27 Hayes, op.cit., pp.382-3.

CHAPTER 2: THE TRIBUNALS: THEIR PROCEDURES AND DECISIONS

1 'Hansard', 346, 1154-5, 26 April 1939.
2 Ibid., 1156.
3 For the next day's full debate, see 'Hansard', 346, 1343-454, 27 April 1939.
4 Ibid., 1156, 26 April 1939.
5 It was decided to combine the duties of labour relations with recruitment and wartime labour control in one Ministry and the formal change was made, despite some opposition from the Ministry of Labour, on 8 September 1939. See H.M.D. Parker, 'Manpower', pp.56-7.
6 See John Rae, 'Conscience and Politics', pp.240-1.
7 National Service (Armed Forces) Act, 1939, section 5.
8 This provision was omitted from the draft of the Military Training Bill but Chamberlain had undertaken in Cabinet to include it if the Labour Party specifically asked for it, PRO, CAB, 25(39)4, 1 May 1939.
9 This, too, was a matter for discussion in Cabinet; the Secretary of State for War, Leslie Hore-Belisha, commented on 3 May 1939 that the moral grounds for a conscientious objection were more important than any legal consideration, PRO, CAB, 26(39)6, 3 May 1939.
10 The quorum necessary to constitute a meeting of a Local Tribunal was to consist of the chairman and two other members.
11 National Service (Armed Forces) Act, 1939, section 5, Miscellaneous Regulations, paras.11-23.
12 By the end of the war the number of Local Tribunals had risen from fifteen to nineteen.
13 CBCO, PC, 'Manchester Guardian', 11 April 1940.
14 PRO, Lab 6, p.148, 1939.
15 Ibid., p.12, 1939.
16 Ibid., p.7, 1939.
17 CBCO, PC II, 'Glasgow Evening Citizen', 27 June 1941.
18 Appendix 3, Table 1, p.144.
19 CBCO, Research and Statistics, memorandum by Judge Wethered OBE, 1942.
20 Ibid.
21 CBCO, PC, unnamed cutting, 4 June 1940.
22 Ibid., memorandum to CBCO Meeting, 5 June 1940.
23 CBCO, Research and Statistics, memorandum by Judge Wethered OBE, 1942.

24 PRO, Lab 6, p.127, 1940.
25 CBCO, PC, 'Manchester Guardian', 12 July 1941.
26 'Hansard', 346, 2097, 4 May 1939.
27 CBCO, ECM, 21, 16 September 1942.
28 Ibid., BM 385, 9 January 1943.
29 Ibid., LTR I-I, 22 November 1939.
30 Ibid., 4 December 1939.
31 Ibid., February 1939.
32 Ibid., 23 April 1940.
33 Ibid., LTR II-I, 8 November 1939.
34 Ibid., 22 November 1939.
35 Ibid., 20 May 1940.
36 Ibid., Research and Statistics, memorandum by Judge Wethered OBE, 1942.
37 Ibid., LTR I-I, 22 November 1939.
38 Ibid., 6 March 1940.
39 PRO, Lab 6, p.142, 1940.
40 CBCO, ATR I-I, 6 December 1939.
41 Ibid., ECM 12, 23 March 1943.
42 Ibid., LTR I-I, 25 June 1940.
43 Ibid., LTR II-I, 8 November 1939.
44 Ibid., 26 April 1940.
45 Ibid., memorandum to CBCO Meeting, 5 June 1943. See Appendix 3, Table 3g, p.152.
46 Ibid. Of a total of 59,192 applications in Local Tribunals in the Second World War, only 6.04 per cent was awarded unconditional exemption. See 'Ministry of Labour and National Service Report 1939-46' and Appendix 3, Table 2, p.145, although the percentage here is lower because the figures include 1945-8 when unconditional exemption was rare. Also see Rae, op.cit., p.132, for a comparison with the First World War.
47 48.52 per cent of applicants were awarded this decision.
48 PRO, Lab 6, p.127, September 1939.
49 Ibid., p.141, 1943.
50 Ibid., p.127, December 1939.
51 Ibid.
52 See Parker, op.cit., pp.128-9 for unevenness of labour distribution in agriculture in 1939.
53 24.82 per cent of objectors were given this decision. This compares with a figure of 35.6 per cent in the Great War when non-combatant service was more popular with Tribunals. See Rae, op.cit., p.132 and pp.117-33.
54 CBCO, LTR II-I, 3 November 1939.
55 Ibid., 23 February 1940.
56 CBCO, BM, 301, 1 August 1941.
57 PRO, Lab 6, p.151, 1940.
60 National Service Act, 1941.
61 20.62 per cent of objectors were given the 'D' decision compared to 17.6 per cent in the 1914-18 war, Rae, op.cit., p.132. This slight increase is interesting and it can perhaps be explained in that there were more applicants in 1939-45 who had grown up in a nationally pacifist atmosphere and who were, therefore, more attracted to registering as conscientious objectors, but found at their Tribunal hearing that their

objections were not as deeply or sincerely held as they had hitherto believed.
62 Appendix 3, Tables 2 and 6a, pp.145 and 156.
63 National Service (Armed Forces) Act, 1939, section 5.
64 National Service Act, 1941.
65 CBCO, 'Yorkshire Post', 10 August 1940.
66 Ibid., 'Law Journal', 17 May 1941.
67 CBCO, ECM, 4, 13 August 1941.
68 Appendix 3, Tables 4 and 6b, pp.154 and 157.

CHAPTER 3: TRIBUNALS IN ACTION: THEIR WORK AND AN ASSESSMENT OF IT

1 Military Training Act (2), 1916.
2 John Rae, 'Conscience and Politics', Chapter 3.
3 John Hughes, The Legal Implications of Conscientious Objection, 1971.
4 National Service (Armed Forces) Act, 1939.
5 'Hansard', 347, 206, 8 May 1939.
6 Ibid., 119.
7 Ibid., 206, 8 May 1939.
8 Ibid., 1427, 27 April 1939.
9 Ibid., 347, 736, 11 May 1939.
10 Ibid., 347, 1301-2, 16 May 1939.
11 Ibid., 347, 729.
12 Ibid., 357, 1622, 22 February 1940.
13 See Appendix 1, pp.138-9.
14 National Service (Armed Forces) Act, 1939, section 5.
15 L.E. White, 'Who Was Who', 1941-50, 1951-60, 1961-70, and 'Who's Who', 1971.
16 CBCO, LTR II-1, South-Eastern LT, 31 January 1940.
17 Ibid., LTR I-7, London LT, 12 October 1939.
18 Ibid., LTR I-2, London LT, 6 August 1940.
19 Ibid., LTR I-1, London LT, 12 October 1939.
20 Ibid.
21 Ibid., 3 April 1940.
22 Ibid., LTR II-1, South-Eastern LT, 29 April 1940.
23 Ibid., LTR I-1, London LT, 22 November 1939.
24 Ibid., 4 December 1939.
25 Ibid., PC II, 'Manchester Evening Chronicle', 9 September 1941.
26 Ibid., 'Harrowgate Herald', 2 July 1941.
27 Ibid., PC, 'Scotsman', 25 April 1940.
28 Ibid., 'News Chronicle', 9 August 1940.
29 Ibid., PC, July-December 1943, 'Northern Daily Telegraph', 22 September 1943.
30 Ibid., Research and Statistics, memorandum by Judge Wethered OBE, 1942.
31 Ibid., LTR I-1, London LT, 19 December 1939.
32 Ibid., Bundle, AT Analysis, December-March, year not dated.
33 See also Denis Hayes, 'Challenge of Conscience', pp.201-2 which supports this view.
34 Ibid., PC II, 'Schoolmaster', 11 September 1941.
35 Ibid., LTR I-1, London LT, 3 April 1940.
36 Ibid., 22 November 1939.

37 'Hansard', 357, 1587-9, 22 February 1940.
38 CBCO, BM, 367, 6 June 1942.
39 Ibid., PC, 'Daily Herald, 18 April 1940.
40 Ibid., 'Daily Mirror', 21 June 1940.
41 'Hansard', 357, 1615, 22 February 1940.
42 Three new Appellate Tribunals were created in 1940. The England and Wales Appellate Tribunal became known as the Southern Appellate Tribunal and had three divisions, the original Tribunal, Division I, and two new Tribunals, Divisions II and III. A Northern Appellate Tribunal was also created.
43 CBCO, BM, 353, 6 June 1942.
44 CBCO, ATR, I-1, England and Wales AT, 18 December 1939.
45 PRO, Lab 6, p.151, National Service Act, 1939, Appeals by Minister to Appellate Tribunal, 1940.
46 CBCO, Research and Statistics, memorandum by Judge Wethered OBE, 1942.
47 Ibid., Bundle, Law Journal, 16 March 1940.
48 Ibid., PC, 'Manchester City News', 14 December 1940.
49 Ibid., 'Daily Mirror', 4 July 1940.
50 Ibid., LTR II-1, South Eastern LT, 8 March 1940.
51 Ibid., BM, 208, 4 December 1940.
52 Ibid., PC II, 'Preston Guardian', 30 April 1941.
53 'Hansard', 357, 1594, 22 February 1940.
54 Ibid., 352, 1007, 19 October 1939.
55 CBCO, LTR II-1, South-Eastern LT, 5 September 1940.

CHAPTER 4: OBJECTORS IN CIVILIAN LIFE: THE UNCONDITIONALLY AND CONDITIONALLY EXEMPTED OBJECTORS

1 National Service (Armed Forces) Act, 1939, section 5.
2 PRO, Lab 6, p.127, 12 December 1939.
3 Ibid., 4 March 1940.
4 Ibid., 15 March 1940.
5 'Hansard', 347, 750, 11 May 1939.
6 Ibid., 1659, 18 May 1939.
7 Ibid., 1665.
8 Ibid., 753, 11 May 1939.
9 Ibid., 754.
10 Some unconditionally exempted objectors chose to work on the land for community service reasons and others, having lost their own jobs because of their views, looked to the land for employment.
11 CBCO, PC, 'Daily Dispatch', 17 February 1940.
12 Ibid., 'Weekly Scotsman', 27 April 1940.
13 Ibid., 'Evening Standard', 31 January 1941.
14 Ibid., 'Farmers' Weekly', 2 May 1941.
15 'Hansard', 362, 215-6, 20 June 1940.
16 See E.C. Urwin, 'Henry Carter C.B.E.: A Memoir', London, 1955.
17 Hayes, 'Challenge of Conscience', pp.208-11.
18 F.A. Lea, 'The Life of John Middleton Murry', London, 1955, C. XXIV. See also Robert Speaight, 'Life of Eric Gill', London, 1966 and D.L. Plowman, 'Bridge into the Future, Letters of Max Plowman', London, 1944.

19 'Hansard', 368, 1101-2, 6 February 1941.
20 CBCO, PC, 'Western Morning News', 27 January 1941.
21 Ibid., 'Worcester News and Times', 5 April 1941.
22 See Denis Hayes, 'Challenge of Conscience', pp.182-200 for an account of their work; a George Medal and a British Empire Medal were among awards going to conscientious objectors in Civil Defence.
23 'Hansard', 362, 1005, 4 July 1940.
24 PRO, Lab 6, p.127, September 1940.
25 Edward Blishen in 'A Cack-handed War', London, 1972, vividly describes his experiences on the land as a conscientious objector during the war.
26 CBCO, PC II, 'Manchester Daily Dispatch', 20 August 1941.
27 Ibid., 'Western Mail', 1 August 1941.
28 Ibid., PC, 'Friend', 9 May 1941.
29 Ibid., ECM, 12, 31 May 1944.
30 'Hansard', 359, 676-78, 11 April 1940.
31 PRO, Lab 6, p.127, December 1939.
32 CBCO, PC I, 'Country Life', 11 May 1940.
33 Ibid., PC II, 'Northampton Chronicle and Echo', 4 September 1941.
34 Ibid., 'Romford Recorder', 18 July 1941.
35 Ibid., 'Daily Mirror', 5 September 1941.
36 Ibid., PC, 'Bath Chronicle and Herald', 3 August 1940.
37 Ibid., 'Wiltshire News', 2 August 1940. Later in the war, agricultural wages substantially improved. See H.M.D. Parker, 'Manpower', pp.429-30.
38 Ibid., PC II, 'Preston Guardian', 2 August 1940.
39 Ibid., PC, 'The Times', 20 May 1941.
40 Ibid., 'Northampton Independent', 18 April 1941.
41 Ibid., PC I, 'Spectator', 7 June 1940.
42 Ibid., PC II, 'Worcester News and Times', 18 August 1941.
43 Ibid., PC, 'Birmingham Evening Dispatch, 13 May 1941.
44 Ibid., PC II, Weekly Review, 14 August 1941.
45 'Hansard', 400, 1525, 8 June 1944.
46 See Chapter 5.
47 CBCO, PC I, 'Yorkshire Observer', 18 April 1940.
48 Hayes, 'Challenge of Conscience', p.38. See also pp.231-6 for an account of the outstanding work done by the FAU during the war years and A. Tegla Davies, 'Friends Ambulance Unit: The Story of the F.A.U. in the Second World War', 1962.
49 CBCO, BM, 420, 9 October 1943.
50 Hayes, op.cit., p.38.
51 CBCO, BM 424, 27 November 1943.
52 Ibid., Paper 001.4, 5 April 1944.
53 'Hansard', 395, 1657-8, 16 December 1943.
54 Ibid., 396, 1120-1, 1 February 1944.
55 CBCO, LTR II, South-Eastern LT, 1 March 1940.
56 Ibid., ECM, 12, 22 October 1941.
57 In January 1942 an Order in Council was issued empowering the Ministry of Labour to direct men to enrol in the Home Guard, but objectors on the conscientious objectors' register were exempted, and those not on the register could appeal on grounds of conscience. See Parker, op.cit., pp.164-5. Nevertheless

there were 59 convictions of people claiming, but not getting exemption on grounds of conscience, Hayes, op.cit., p.389.
58 CBCO, PC II, 'Cooperative News', 20 September 1941.
59 Ibid., PC, 'Midlands Counties Tribune', 2 August 1940.
60 Ibid., 'Manchester City News', 15 March 1941.
61 Ibid., 'West Lancashire Evening Gazette', 4 April 1940.
62 Ibid.,Paper 001.5, 22 September 1942.
63 Ibid., ECM, 11, 18 March 1942.
64 See David Morris, 'China Changed My Mind', London, 1948, for an account of one objector in the FAU who was sent to China and as a result of his experiences there, recanted his pacifism.
65 PRO, Lab 6, p.229, 1943-5.
66 CBCO, BM, Report, 17 September 1940.
67 See chapters 14 and 15 of Hayes, op.cit., for accounts of the work of these organisations.
68 CBCO, PC, 'Manchester Guardian, 10 April 1941.
69 Ibid., PC II, 'Sheffield Telegraph', 20 September 1941.
70 Ibid., PC, 'Daily Express', 30 April 1941.
71 'Hansard', 379, 1868, 14 May 1941.
72 Kenneth Mellanby's work, and the part conscientious objectors played in it, is described in his own book, 'Human Guinea-Pigs', London, 1945.
73 See above p. 26 et seq.
74 PRO, Lab 6, p.128, February 1940.
75 Ibid., p.165, 16 October 1940.
76 Ibid., p.189, 21 May 1941.
77 CBCO, PC, 'Evening Chronicle', 29 June 1940.
78 Ibid., 'Evening Standard', 21 March 1940.
79 CBCO, PC, 'Friend', 26 July 1940. Unemployment in the population was just over 1 million in 1939 dropping quickly in 1940 to reach the lowest point in 1944: 400,000. See Parker, op.cit., p.481.
80 Hayes, op.cit., pp.277-81.
81 'Hansard', 370, 603-5, 26 March 1941.
82 Ibid., 626-9.
83 Ibid., 880, 1 April 1941.
84 Ibid., 843.
85 Ibid., 884.
86 Ibid., 889-90.
87 Ibid., 890-2.
88 PRO, HO 186, p.616, 23 January 1941.
89 Home Security Circular, no. 169/1941.
90 Ibid. and see chapter 5 for the treatment of conscientious objectors by local authorities in the early years of the war.
91 PRO, HO 186, p.181, Ministry of Labour Circular, 123/26.
92 Ibid., p.616, National Service Bill, Notes on Clauses.
93 CBCO, Paper 104, Circular 163/1943.
94 Hayes, op.cit., p.389

CHAPTER 5: PUBLIC EMPLOYERS AND THEIR ATTITUDE TO THE EMPLOYMENT OF CONSCIENTIOUS OBJECTORS

1 CBCO, LAD, West Penwith, 3 October 1940.
2 Ibid., Long Eaton, 25 July 1940.
3 Ibid., Northwich, 24 February 1940.
4 Ibid., PC I, 'Manchester Daily Dispatch', 20 May 1940.
5 Ibid., 'Sunday Pictorial', 14 July 1940.
6 Ibid., LAD, Cheadle, not dated.
7 Ibid., Lytham St Annes, April 1940.
8 Ibid., Essex, 21 November 1940.
9 Ibid., Merton, not dated.
10 Ibid., Hamilton, 9 February 1944.
11 Ibid.
12 Ibid., Portsmouth, 12 June 1940.
13 Ibid., Aldershot, 8 September 1940.
14 Ibid., PC I, 'Hertfordshire Mercury', 2 August 1940.
15 Ibid., PC II, 'Evening Citizen', 6 August 1941, 'Reynolds News', 10 August 1941.
16 Ibid., LAD, Acton, 19 July 1940.
17 Ibid., Bournemouth, 25 September 1940.
18 Ibid., Berkshire, not dated.
19 Ibid., Gloucester, 29 June 1940.
20 Ibid., Leamington, not dated.
21 Ibid., Herefordshire, 29 July 1940.
22 Ibid., Esher, not dated.
23 Ibid., Welwyn Garden City, not dated.
24 Ibid., Wimbledon, 30 January 1942.
25 Ibid., 'Manchester Guardian', 1 August 1941.
26 Ibid., PC II, 'Hampshire Herald', 29 August 1941.
27 Ibid.
28 Ibid., PC I, 'Sunday Comet', 6 July 1940.
29 Ibid., PC II, 'Widnes Weekly News', 15 August 1941.
30 Ibid., LAD, Barnet, 18 January 1941.
31 'Hansard', 370, 284, 20 March 1941.
32 Ibid., 363, 1371, 1 August 1940.
33 Ibid., 364, 934, 15 August 1940.
34 Ibid., 371, 1602-3, 22 May 1941.
35 Ibid., 2002, 29 May 1941.
36 Ibid., 376, 1715, 11 December 1941.
37 CBCO, PC, 'Manchester Guardian', 13 August 1940.
38 Ibid., 14 August 1940.
39 Ibid., 23 August 1940.
40 Ibid., PC II, 'Carlisle Journal', 25 April 1941.
41 Ibid., PC, 'Solicitor', 9 November 1940.
42 Ibid.
43 Ibid., 'Manchester Guardian', 5 July 1940.
44 Ibid., 'County Advertiser and Herald', 26 September 1940, 'East Kent Gazette', 14 December 1940.
45 Ibid., PC II, 'Hampshire Herald', 29 August 1941.
46 Ibid., 'Halifax Daily Courier and Guardian', 17 July 1941.
47 Ibid., LAD, Leyton, 4 May 1940.
48 Denis Hayes, 'Challenge of Conscience', p.206.
49 CBCO, LAD, Chadderton, 14 May 1940.

50 Ibid., PC II, 'Bolton Evening News', 1 August 1940.
51 See Table 1, p.69, and Appendix 2, pp.140-2.
52 CBCO, PC I, 'Hemel Hempstead Gazette', not dated.
53 Ibid., 'Daily Sketch', 20 April 1940.
54 'Hansard', 359, 581-2, 10 April 1940.
55 Ibid., 364, 1485, 22 August 1940.
56 Ibid., 365, 1238-9, 5 November 1940.
57 Ibid., 372, 1483, 3 July 1941.
58 CBCO, LAD, Fife County Council, 16 October 1940.
59 Ibid., PC II, 'Northamptonshire Evening Telegraph', 1 August 1941.
60 Ibid., 'Manchester Guardian', 22 July 1941.
61 Ibid., LAD, North Riding, 13 November 1940.
62 Ibid., Material for 'A Plea for Tolerance', 26 July 1940.
63 'Hansard', 364, 1464-5, 22 August 1940.
64 PRO, Ministry of Health Circular 2034, May 1940.
65 CBCO, PC, 'Eduation', 19 August 1940.
66 Ibid., 'Manchester Guardian', 23 August 1940.
67 Ibid., 'Teacher's World', 14 August 1940.
68 Ibid., LAD, 'Kent Messenger', 27 February 1942.
69 'Hansard', 346, 2100, 4 May 1939.
70 CBCO, LAD, Circular 1522.
71 Ibid., Papers, 104, Circular 169/1941, 1 August 1941.
72 CBCO, PC II, 'Coulsdon and Purley Times', 29 August 1941.
73 Ibid., 'Belfast News Letter', 22 July 1941.
74 Ibid., 'Hornsey Journal', 8 August 1941.
75 Ibid., 'Enfield Gazette and Observer', 8 August 1941.
76 Ibid., 'Barnet Press', 30 August 1941.
77 'Hansard', 364, 1497-8, 22 August 1940.
78 CBCO, PC I, 'Essex County Telegraph', 13 July 1940.
79 'Hansard', 363, 36-7, 16 July 1940.
80 CBCO, PC, 'Sunday Dispatch', 22 September 1940.
81 Ibid., 'Manchester Guardian', 20 December 1940.
82 Ibid., 'Daily Herald', 20 December 1940.
83 'Hansard', 370, 284, 20 March 1941. It was on this occasion that Churchill made his often quoted remark that persecution was 'odious to the British people'.
84 CBCO, PC II, 'Western Mail', 20 August 1941.
85 Ibid., 2 September 1941.
86 Ibid., 26 August 1941.
87 Ibid., LAD, Long Eaton, 25 July 1940.
88 Ibid., PC I, 'Coventry Herald', 20 July 1940.
89 Ibid., 'Star', Sheffield, 16 July 1940.
90 Ibid., 'Evening Standard', 21 June 1940.
91 Ibid., PC II, 'Evening Standard', 30 June 1941.
92 The resolution was passed; for details see the Report of the 72nd Annual Trades Union Congress 1940 kept at Congress House.
93 CBCO, PCI1, 'Sunday Chronicle' and 'Sunday Referee', 28 April 1940.
94 Ibid., Material for 'A Plea for Tolerance', 16 August 1940.
95 Ibid., PC II, 'Freethinker', 26 May 1940.
96 Ibid., PC, 'World Review', November 1940.
97 Ibid., 'Newcastle Evening Chronicle', 5 August 1940.
98 Ibid., 'Barnet Press', 18 January 1941.

CHAPTER 6: CONSCIENTIOUS OBJECTORS IN THE ARMED FORCES AND IN PRISONS

1 See above p.24 et seq.
2 'Hansard', 353, 522, 14 November 1939.
3 Ibid., 365, 1607, 12 November 1940.
4 CBCO, PC, 'Berwick Advertiser', 15 February 1940.
5 Ibid., 'Daily Dispatch', 25 April 1940.
6 PRO, Lab 6, P.147, 1939.
7 Ibid., February 1940.
8 Ibid., p.126, April 1940.
9 Denis Hayes, 'Challenge of Conscience', pp.123-4.
10 'Hansard', 361, 1114, 11 June 1940.
11 CBCO, PC II, 'The Times', 6 May 1941.
12 Ibid., 'Northampton Independent', 8 August 1941.
13 PRO, WO 166, P.5831, War Diaries, 12 June 1940.
14 Ibid., 31 July 1940.
15 Ibid., 21 September 1940.
16 CBCO Bundle, 'Manchester Evening News', 9 September 1960.
17 Ibid., PC II, 'Empire News', 3 August 1941.
18 CBCO Papers, 699, Misc., 7 April 1942.
19 The War Office confirmed that no conscientious objector was to receive promotion on the grounds that in an emergency an officer's duty would be to lead his men in action, which a conscientious objector quite obviously could not be expected to do. This policy was retained throughout the war, despite appeals from officers to promote conscientious objectors who had performed particularly gallant work in the 6th Airborne Division. PRO, WO 32, P.9432.
20 Hayes, op.cit., p.131. See also War Diaries of the 1st Co. N.C.C. Pioneer Corps, PRO, WO 166 P.5831.
21 'Hansard', 369, 745, 4 March 1941; 1127-8, 11 March 1941.
22 CBCO Bundle, 24 October 1939.
23 Work in the NCC could involve loading and off loading foodstuffs or, perhaps, trench digging for water pipes, PRO, WO 166, P.5831.
24 'Hansard', 365, 1507, 12 November 1940.
25 Hayes, op.cit., pp.127-30.
26 CBCO Papers, Misc., 'Sunday Graphic' and 'Sunday News', 25 June 1944.
27 Hayes, op.cit., p.8.
28 'Hansard', 347, 1681-2, 18 May 1939.
29 Ibid., 1688.
30 Ibid., 1689-90.
31 Ibid.
32 PRO, Lab 6, P.14, 5 July 1941.
33 'Hansard', 347, 1695-6, 18 May 1939.
34 Ibid., Lords, 113, 225, 25 May 1939.
35 CBCO, PC II, 'New Statesman', 12 July 1941.
36 'Hansard', 359, 773-775, 16 April 1940.
37 CBCO, PC II, 'New Statesman', 12 July 1941.
38 Ibid., 'New Leader', 9 August 1941.
39 'Hansard', 359, 773-5, 16 April 1940. See also for details of the case, Hayes, op.cit., pp.78-80.

40 CBCO Papers 303, 30 May 1940.
41 Ibid., 307, not dated.
42 Ibid., 19 August 1941.
43 PRO, Lab 6, P.14, 3 October 1942.
44 Ibid., February 1943. See also Hayes, op.cit., pp.106-11.
45 Hayes, op.cit., p.118.
46 Appendix 3, Table 8, p.158.
47 See below p.93 et seq.
48 'Hansard', 376, 1422-3, 9 December 1941.
49 Hayes, op.cit., p.167.
50 CBCO Papers 406, 1943.
51 Ibid.
52 Ibid., 407, 1961.
53 'Hansard', 360, 1019, 7 May 1940.
54 PRO, WO 32, P.14529, 12 February 1940.
55 'Hansard', 371, 579, 1 May 1941.
56 Ibid., 377, 1379, 10 February 1942.
57 CBCO Papers, 000-002, 001, 10 February 1941.
58 CBCO, ECM, 11 August 1943.
59 Examples of such incidents occur in Hayes, op.cit., pp.88, 117, 354-5.
60 'Hansard', 373, 1246-7, 29 July 1941.
61 Ibid., 371, 325-6, 29 April 1941.
62 John Rae, 'Conscience and Politics', pp.158-61.
63 'Hansard', 358, 982, 7 March 1940.
64 Hayes, op.cit., pp.91-100.
65 'Hansard', 365, 825, 17 October 1940.
66 Ibid., 934-5, 22 October 1940.
67 Ibid., 1177-8, 5 November 1940.
68 Ibid., 1979, 20 November 1940.
69 Ibid., 368, 404-5, 28 January 1941.
70 Ibid., 369, 346-7, 25 February 1941.
71 Ibid., 377.
72 Ibid., 371, 677, 6 May 1941.
73 Hayes, op.cit., p.98.
74 'Hansard', 378, 913-4, 10 March 1942.
75 Ibid., 379, 1731-2, 13 May 1942.
76 National Service (Armed Forces) Act, 1939, section 3(5).
77 Hayes, op.cit., p.144.
78 Sybil Morrison, 'I Renounce War: The Story of the Peace Pledge Union', London, 1962, pp.52-3.
79 Interview with ex-CO, May 1977.
80 CBCO, Home Services Committee, Prisons WO 1942-8, June 1943.
81 Ibid., March 1943.
82 Ibid., Letter to Quaker Ministers from Robert Davis, Secretary, 15 October 1941.
83 See below p.97.
84 'Hansard', 373, 1522-3, 31 July 1941.
85 Interview with ex-CO, May 1977.
86 'Hansard', 374, 1132, 9 October 1941.
87 Ibid., 383, 179, 9 September 1942.
88 Ibid., 390, 378, 3 June 1943.
89 CBCO, PC II, 'New Leader', 27 September 1941.
90 CBCO, ECM, 4 February 1942.

91 Ibid., PC II, 'Aberdeen Evening Express', 4 September 1941. See Alfred Hassler, 'Diary of a Self-made Convict', 1955, for an account of the experiences of a conscientious objector in a United States prison.

CHAPTER 7: THE PEACE PLEDGE UNION AND THE CENTRAL BOARD FOR CONSCIENTIOUS OBJECTORS

1 Denis Hayes, 'Challenge of Conscience', p.74.
2 'Peace News', 8 September 1939.
3 Martin Ceadel, 'Pacifism in Britain 1914-45', p.318.
4 'Hansard', 357, 1504-6, 22 February 1940. In fact, it had been an 'official decision' not to try to secure young men at the moment of registration at Labour Exchanges by distribution of pacifist literature. A document was left explaining registrants' rights and open-air meetings were held in the vicinity. See Sybil Morrison, 'I Renounce War', p.49.
5 Morrison, op.cit., p.44. Frances Partridge in 'A Pacifist's War', 1977, recalls her family's rejection by some of the village community where she spent the war.
6 Morrison, op.cit., p.44.
7 Ibid., p.53.
8 F.A. Lea, 'Life of John Middleton Murry', p.280.
9 Ibid., p.282.
10 Ibid., p.284.
11 Ibid., p.310.
12 Ceadel, op.cit., p.310.
13 CBCO, PC II, 'Brighton Gazette', 6 September 1941.
14 'Hansard', 358, 1366-7, 6 June 1940.
15 Ibid.
16 Ibid., 361, 986-7, 6 June 1940.
17 These were Alex Wood (PPU Chairman), Stuart Morris (General Secretary), Maurice Rowntree (Treasurer) and John Barclay, with two Group Leaders, Ronald Smith and Sidney Todd, Morrison, op.cit., pp.46-9.
18 CBCO, PC, 'Birmingham Gazette', 10 May 1940.
19 Ibid., 'The Times', 12 June 1940.
20 Ibid., 'Daily Mail', 25 May 1940.
21 'Hansard', 378, 2137-8, 26 March 1942.
22 Ibid., 392, 892, 13 October 1943.
23 Ceadel, op.cit., pp.294-315.
24 Hayes, op.cit., p.vii.
25 Ibid., p.74.
26 'Daily Mail', 5 April 1940.
27 CBCO, BM, 197, 13 November 1940.
28 Ibid., 2, 11 January 1940.
29 Ibid., 3, 29 January 1940.
30 Ibid., ECM, 16 April 1941.
31 For his utilitarian pacifism, see Arthur Ponsonby, 'Now Is The Time: An Appeal for Peace', London, 1925.
32 CBCO, BM, 23, 22 February 1940.
33 Ibid., 11, 8 February 1940.
34 Ibid., 27 September 1941.

35 Ibid., 24 September 1941.
36 Hayes, op.cit., p.x.
37 CBCO, Publications I.
38 Hayes, op.cit., p.x.
39 CBCO, PC, 'Bournemouth Daily Echo', 25 May 1940.
40 Ibid., Papers 699, 1 October 1942.
41 Ibid., ECM, 23 October 1942.
42 Ibid., Papers 699, 16 October 1942.
43 Hayes, op.cit., p.259.
44 CBCO, BM, 301, 27 September 1941.
45 Hayes, op.cit., pp.261-3.
46 PRO, Lab 186, P.1041, 1942.
47 Ibid.
48 Women's conscription for military service was introduced at the end of 1941 but it was limited to young single women only. This is discussed on pp.110-11.
49 CBCO, ECM, 19 April 1944.
50 Ibid., 26 April 1944.
51 Hayes, op.cit., p.389.
52 CBCO, BM, 228, 22 January 1941.
53 Hayes, op.cit., p.299.
54 PRO, HO 186, P.822, 3 December 1941.
55 Ibid., 14 January 1942.
56 Ibid., 16 January 1942.
57 Hayes, op.cit., pp.318-24.
58 Ibid., p.389. Kathleen Lonsdale, a Fellow of the Royal Society and later a DBE, was imprisoned for her refusal to firewatch under compulsion. See Morrison, op.cit., p.54.
59 'Hansard', 378, 818-67, 5 March 1941.
60 CBCO, Papers 000, 9 April 1942.
61 Appendix 3, Tables 2 and 6a, pp.145 and 156.
62 CBCO, BM, 293, 1 August 1941.
63 Ibid., 305, 25 October 1941.
64 See chapter 5.
65 PRO, Lab 6, P.172, 5 July 1940.
66 Ibid.
67 Ibid.
68 Ibid.
69 'Hansard', 368, 1054, 6 February 1941.
70 CBCO, BM, 285, 1054, 6 February 1941.
71 Hayes, op.cit., p.xi.
72 CBCO, BM, 231, 22 January 1941.
73 Ibid., 329, 17 January 1942.

CONCLUSION

1 Appendix 3, Table 6a, p.156.
2 Denis Hayes, 'The Challenge of Conscience', p.27.
3 See above p.54.
4 Keith Mellanby, 'Human Guinea-Pigs', 1945.
5 Constance Braithwaite, 'A Short History of the Legal Position of Conscientious Objectors', written 1956, duplicated and re-issued by Friends' Peace Committee, May 1964.

6 Hayes, op.cit., p.207.
7 Ibid.
8 The PPU is still active, as is the CBCO which, despite the absence of conscription, still helps members of the Armed Forces who develop a conscientious objection while in service.
9 Hayes, op.cit., pp.342-59.
10 Braithwaite, op.cit., pp.7 and 10.
11 See Alan Bullock, 'The Life and Times of Ernest Bevin', vol.1, 1960.
12 Ibid.
13 D. Prasad and T. Smythe, 'Conscription: A World Survey', 1968.
14 Greece, now a member of the European Community, still imprisons Jehovah's Witnesses who refuse 'alternative service'.

MEMBERSHIP OF APPELLATE AND LOCAL TRIBUNALS, 4 APRIL 1940

APPELLATE TRIBUNALS

England and Wales

*Rt Hon. H.A.L. Fisher
Hon. Sir Leonard W.J. Costello
Sir Arthur Pugh

Scotland

*Rt Hon. Lord Elphinstone
T. Barron Esq.
Alexander Maitland Esq.

LOCAL TRIBUNALS

London

*Judge Hargreaves
Sir James Baillie
Alderman S.H. Marshall
Sir Edmund Phipps
A.B. Swales Esq.

South-Eastern

*Judge Davis
Sir Reginald Kennedy-Cox
Sir Walter Kinnear
A.N. McConell Esq.
Dr G. Senter
E.G. Witcher Esq.

Southern

*Judge Drucquer
C.H. Davies Esq.
Sir Cyril Norwood
John Simonds Esq.
One vacancy

East-Anglian

*Judge Campbell
J.H. Clapham Esq.
W. Holmes Esq.
Alderman James Nutter
Alderman J.H. Stadden

South-Western

*Judge Wethered
Dr C. Bailey
C.P. Brown Esq.
A.L. Hobhouse Esq.
Alderman F. Sheppard

Midlands

*Judge Longson
A.H. Gibbard Esq.
George Trevelyan Lee Esq.
Cllr E. Purser
Professor J.G. Smith

North-Eastern

*Judge Stewart
E.C. Behrens Esq.
H. Dodgson Esq.
D. McCandish Esq.
Herbert Woodhouse Esq.

North-Western

*Judge Burgis
Alderman C. Aveling
L.F. Behrens Esq.
A. Kerr Esq.
Sir Miles Ewart Mitchell
A. Roberts Esq.

Cumberland and Westmoreland

*Judge Peel
Frank A. Carr Esq.
R.D. McCowan Esq.
Charles H. Roberts Esq.
A. Stephenson Esq.

Northumberland and Durham

*Judge Richardson
Professor James F. Duff
Alderman F. Nicholson
Sir Luke Thompson
One vacancy

North Wales

*Judge Sir T. Artemus Jones
Alderman T. Lloyd Williams
Alderman Sir Evan Jones
Cllr R. Owen
Alderman H.R. Thomas

South Wales

*Judge Frank Davies
James Evans Esq.
O. Harris Esq.
Sir L. Twiston Davies
J.H. Williams Esq.

South-East Scotland

*C.H. Brown Esq. KC
W.H. Brist Esq.
W. Nellies Esq.
E.M. Wedderburn Esq.
One vacancy

South-West Scotland

*Sir Archibald Campbell Black
Sir Robert Bruce
Robert Bryce Walker Esq.
W. Coriner Esq.
One vacancy

North Scotland

*Robert H. Maconochie Esq.
P. Campbell Esq.
Provost Hugh Mackenzie
Dr G.G. Middleton
F.O. Stuart Esq.

North-East Scotland

*Sir George Morton
G. Duncan Esq.
Principal W. Hamilton Fyfe
G.R. McIntosh Esq.
Provost J.C. Rankin

* Chairman

COUNTY AND CITY COUNCIL DECISIONS ON EMPLOYMENT OF CONSCIENTIOUS OBJECTORS

County Council Decisions (England and Wales)

Councils dismissing objectors

Buckinghamshire
Caernarvonshire
Cambridgeshire
Cardiganshire
Cheshire
City of London
Derbyshire
Devonshire
Dorsetshire
Hertfordshire
Isle of Ely
Lincolnshire (Kesteven)
Northumberland
Pembrokeshire
Salop
Suffolk (West)
Warwickshire
Worcester
Yorkshire (West Riding)

TOTAL: 19

Councils dismissing objectors for the duration of the war

Anglesey
Berkshire
Kent
Middlesex
Surrey
Yorkshire (North Riding)

TOTAL: 6

Councils retaining their objectors

Essex
Gloucestershire
London County Council
Montgomeryshire
Somerset

TOTAL: 5

Councils not accounted for

Brecknockshire
Carmarthenshire
Cornwall
Cumberland
Durham
Glamorgan
Huntingdonshire
Isle of Wight
Lincolnshire (Holland)
Lincolnshire (Lindsey)
Merionethshire
Monmouthshire
Norfolk
Northamptonshire
Northamptonshire (Soke of Peterborough)
Radnorshire
Rutland
Southampton
Staffordshire
Suffolk (East)

Sussex (East)
Sussex (West)
Westmoreland
Wiltshire
Yorkshire (East Riding)

TOTAL: 25

Councils retaining objectors on soldiers' pay

Bedfordshire
Denbighshire
Flintshire
Herefordshire
Lancashire
Leicestershire
Nottinghamshire
Oxfordshire

TOTAL: 8

SUM TOTAL: 63

County and City Boroughs in England and Wales

Councils dismissing objectors

Blackpool
Brighton
Burton-on-Trent
Bury
Cardiff
Carlisle
Chichester
Croydon
Darlington
Doncaster
Dudley
Eastbourne
Grimsby
Huddersfield
Lincoln
Manchester
Middlesborough
Nottingham
Oldham
Oxford
Plymouth
Portsmouth
Salford
Southport
Stockport
Stoke-on-Trent
Wakefield
Walsall
West Ham
West Hartlepool
Worcester
York

TOTAL: 32

Councils dismissing objectors for the duration of the war

Birmingham
Bootle
Bradford
Derby
East Ham
Norwich
Reading
Smethwick
Southampton
Swansea
Warrington
Wells
West Bromwich
Wolverhampton

TOTAL: 14

Councils retaining their objectors

Birkenhead
Bournemouth
Bristol
Chester
Coventry
Dewsbury
Durham
Gloucester
Hastings
Hereford
Ipswich
Leeds
Liverpool
Merthyr Tydvil
Newcastle-upon-Tyne
Rochdale
Rotherham
Sheffield
Southend-on-Sea

TOTAL: 19

Councils not accounted for

Bangor
Barnsley
Barrow-in-Furness
Blackburn
Burnley
Exeter
Gateshead
Great Yarmouth
Halifax
Kingston-upon-Hull
Lancaster
Newport (Monmouthshire)
Peterborough
Preston
Ripon
St Albans
St Helens
Salisbury
South Shields
Sunderland
Truro
Tynemouth
Wallasey
Wigan

TOTAL: 24

Councils retaining objectors on soldiers' pay

Bath
Bolton
Canterbury
Leicester
Northampton
Rochester

TOTAL: 6

SUM TOTAL: 95

Appendix 3

STATISTICS

Source: Ministry of Labour and National Service, PRO, Lab 6, Pieces 405 and 337.

The embargo on these hitherto closed filed was lifted just in time to include the statistics they contain in this book. They are virtually identical to those in Denis Hayes's 'Challenge of Conscience' whose author obtained them from the Ministry of Labour after the war but before the files were closed. In most cases they run to the end of 1948 when previous National Service Acts were superseded by the National Service Acts, 1948, which took effect on 1 January 1949. The Acts merely consolidated and reissued previous regulations.

TABLE 1 Number of men registering provisionally in the Register of Conscientious Objectors from 1939 to 1945 inclusive

Registration date	No. registering provisionally as COs in each period	Total registrations in each period	Percentage of provisional registrations
Military Training Act 1939			
1 3 June 1939	4,392	240,757	1.8
National Service Acts 1939-42			
2 21 Oct. 1939	5,073	230,009	2.2
3 9 Dec. 1939	5,490	256,300	2.1
4 17 Feb. 1940	5,638	278,289	2.0
5 9 Mar. 1940	5,803	346,731	1.6
6 6 Apr. 1940	4,772	335,909	1.4
7 27 Apr. 1940	4,218	336,894	1.2
8 25 May 1940	3,684	348,991	1.05
9 15 June 1940	2,387	307,858	0.77
10 22 June 1940	2,451	355,105	0.69
11 6 July 1940	1,898	330,456	0.57
12 13 July 1940	1,752	342,367	0.51
13 20 July 1940	1,669	331,030	0.5
14 27 July 1940	2,192	380,087	0.57
15 9 Nov. 1940) 16 16 Nov. 1940)	2,173	407,302	0.53
17 11 Jan. 1941) 18 18 Jan. 1941)	1,658	366,684	0.45
19 22 Feb. 1941	1,674	291,143	0.57
20 12 Apr. 1941	1,342	319,456	0.42
21 17 May 1941	1,176	323,881	0.36
22 31 May 1941	1,170	306,907	0.38
23 21 June 1941	558	152,107	0.36
24 12 July 1941	665	142,671	0.47
25 6 Sep. 1941	696	156,465	0.44
26 13 Dec. 1941	657	162,926	0.4
27 18 Apr. 1942	608	157,654	0.38
28 15 Aug. 1942	539	158,000	0.34
29 7 Nov. 1942	310	83,457	0.37
30 9 Jan. 1943	481	159,690	0.30
31 3 Apr. 1943	301	83,867	0.35
32 19 June 1943	267	79,864	0.33
33 18 Sep. 1943	187	70,810	0.26
34 11 Dec. 1943	173	71,033	0.24
35 4 Mar. 1944	176	71,920	0.24
36 3 June 1944	181	69,420	0.26
37 2 Sep. 1944	148	67,847	0.22
38 2 Dec. 1944	155	76,563	0.2
39 3 Mar. 1945	176	84,017	0.21
40 9 June 1945	157	72,028	0.21
41 1 Sep. 1945	130	68,553	0.18
42 1 Dec. 1945	159	75,343	0.2

TABLE 2 Analysis of decisions of Local Tribunals for registration of Conscientious Objectors up to 31 December 1948

Tribunal	Registered as COs unconditionally		Registered as COs on condition that they undertook civil work		Registered as COs but liable to be called for non-combatant duties in HM Forces		Removed from register of COs		Total
	No.	%	No.	%	No.	%	No.	%	
London No 1	145	1.1	3,083	23.2	4,852	36.4	5,239	39.3	13,319
London No 2	75	3.0	667	29.0	527	23.0	1,025	45.0	2,294
South-East	221	4.0	2,106	37.3	1,434	25.4	1,880	33.3	5,641
Southern	45	1.3	1,237	37.1	1,039	31.2	1,012	30.4	3,333
E Anglia	275	10.2	1,395	51.8	693	25.8	329	12.2	2,692
S Western	567	11.1	2,770	53.9	1,204	23.5	589	11.5	5,130
Midlands	98	1.5	4,205	65.1	995	15.4	1,165	18.0	6,463
N Midlands	4	0.2	942	43.4	581	26.7	645	29.7	2,172
N Eastern	80	2.5	1,057	33.2	1,110	34.9	934	29.4	3,181
N Western	373	7.4	1,752	34.4	1,670	32.8	1,293	25.4	5,088
Cumberland and Westmoreland	30	6.5	99	21.4	218	47.2	115	24.9	462
Northumberland and Durham	114	9.9	431	37.6	288	25.1	315	27.4	1,148
SE Scotland	96	6.5	395	27.0	536	36.7	436	29.8	1,463
SW Scotland No.1	449	14.7	568	18.6	377	12.3	1,659	54.4	3,053
SW Scotland No.2	11	0.8	376	28.7	196	15.0	727	55.5	1,310
N Scotland	5	3.7	21	15.3	48	35.0	63	46.0	137
NE Scotland	19	3.9	209	42.7	158	32.3	103	21.1	489
S Wales	252	8.3	1,356	44.6	907	29.8	526	17.3	3,041
N Wales	78	4.1	969	51.4	398	21.1	440	23.4	1,885
Men	2,868	4.7	22,949	37.5	17,193	28.1	18,217	29.7	61,227
Women	69	6.4	689	64.1	38	3.5	278	26.0	1,074
Cumulative total	2,937	4.7	23,638	37.9	17,231	27.7	18,495	29.7	62,301

TABLE 3(a) 1939 Orders made by Tribunals on applications for registration in register of Conscientious Objectors under Section 5 of the National Service (Armed Forces) Act 1939 (Men)*

Tribunal	A	B	C	D
Local Tribunals:				
London No.1	45	442	412	228
London No.2	-	-	-	-
South Eastern	64	239	75	30
Southern	-	-	-	-
East Anglian	122	101	92	25
South Western	149	203	44	14
Midlands	21	346	42	72
North Midlands	-	-	-	-
North Eastern	32	137	158	106
North Western	81	132	42	106
Cumberland and Westmoreland	3	18	11	5
Northumberland and Durham	26	99	34	34
SE Scotland	16	40	34	23
SW Scotland No.1	59	86	79	137
SW Scotland No.2	-	-	-	-
N Scotland	1	4	6	5
NE Scotland	3	8	10	6
S Wales	56	207	75	45
N Wales	21	102	16	13
All Local Tribunals	669	2,164	1,130	849
Appellate Tribunal: (Decisions given on appeal)				
South England No.1 Division	1	36	51	42
South England No.2 Division	-	-	-	-
South England No.3 Division	-	-	-	-
North England Division	-	-	-	-
Wales Division	-	-	-	-
Scotland Division	-	-	-	-
All divisions of Appellate Tribunal	1	36	51	42

* Key to Decisions:

A Registration in COs register unconditionally.

B Registration in COs register conditionally upon undertaking civilian work or training.

C Registration in COs register as person liable to be called up but to be employed only in non-combatant duties.

D Name to be removed from COs register.

TABLE 3(b) 1940 Orders made by Tribunals on applications for registration in register of Conscientious Objectors under Section 5 of the National Service (Armed Forces) Act 1939 (Men)*

Tribunal	A	B	C	D
Local Tribunals:				
London No.1	12	793	2,388	1,884
London No.2	75	571	419	842
South Eastern	136	1,060	716	462
Southern	44	675	565	514
East Anglian	137	486	359	105
South Western	149	1,272	712	284
Midlands	42	2,286	596	694
North Midlands	2	343	203	155
North Eastern	29	385	610	439
North Western	217	741	924	653
Cumberland and Westmoreland	19	63	174	96
Northumberland and Durham	49	199	164	114
SE Scotland	59	223	286	153
SW Scotland No.1	211	301	142	877
SW Scotland No.2	8	167	82	185
N Scotland	2	6	21	28
NE Scotland	9	81	65	37
S Wales	57	470	516	234
N Wales	49	621	257	95
All Local Tribunals	1,540	10,743	9,199	8,469
Appellate Tribunal: (Decisions given on appeal)				
South England No.1 Division	27	836	1,095	703
South England No.2 Division	1	299	314	258
South England No.3 Division	-	-	-	-
North England Division	-	29	18	20
Wales Division	-	-	-	-
Scotland Division	29	233	188	347
All divisions of Appellate Tribunal	57	1,397	1,615	1,328

* Key to Decisions:

A Registration in COs register unconditionally.
B Registration in COs register conditionally upon undertaking civilian work or training.
C Registration in COs register as person liable to be called up but to be employed only in non-combatant duties.
D Name to be removed from COs register.

TABLE 3(c) 1941 Orders made by Tribunals on applications for registration in register of Conscientious Objectors under Section 5 of the National Service (Armed Forces) Act 1939 (Men)*

Tribunal	A	B	C	D
Local Tribunals:				
London No.1	26	726	999	1,231
London No.2	-	96	108	183
South Eastern	1	213	320	462
Southern	1	188	204	177
East Anglian	6	342	123	64
South Western	19	600	169	111
Midlands	2	647	207	126
North Midlands	2	276	240	240
North Eastern	7	189	179	153
North Western	45	373	343	200
Cumberland and Westmoreland	6	13	28	13
Northumberland and Durham	13	57	39	53
SE Scotland	9	54	122	99
SW Scotland No.1	40	72	64	225
SW Scotland No.2	2	114	74	289
N Scotland	1	7	12	16
NE Scotland	1	97	44	20
S Wales	75	277	123	80
N Wales	5	134	70	192
All Local Tribunals	261	4,475	3,468	3,934
Appellate Tribunal: (Decisions given on appeal)				
South England No.1 Division	31	1,304	749	738
South England No.2 Division	20	902	624	641
South England No.3 Division	6	335	310	228
North England Division	88	549	409	453
Wales Division	3	181	81	58
Scotland Division	3	334	251	459
All divisions of Appellate Tribunal	151	3,605	2,427	2,577

* Key to Decisions:

A Registration in COs register unconditionally.
B Registration in COs register conditionally upon undertaking civilian work or training.
C Registration in COs register as person liable to be called up but to be employed only in non-combatant duties.
D Name to be removed from COs register.

TABLE 3(d) 1942 Orders made by Tribunals on applications for registration in register of Conscientious Objectors under Section 5 of the National Service (Armed Forces) Act 1939 (Men)*

Tribunal	A	B	C	D
Local Tribunals:				
London No.1	14	391	503	764
London No.2	-	-	-	-
South Eastern	4	334	235	182
Southern	-	144	133	159
East Anglian	1	260	74	67
South Western	9	318	127	75
Midlands	6	348	71	101
North Midlands	-	162	80	126
North Eastern	7	136	95	110
North Western	11	211	201	157
Cumberland and Whstmoreland	2	5	5	1
Northumberland and Durham	13	20	15	50
SE Scotland	3	29	47	68
SW Scotland No.1	52	22	20	107
SW Scotland No.2	1	46	20	134
N Scotland	-	4	8	11
NE Scotland	1	5	17	23
S Wales	34	164	106	74
N Wales	1	59	36	72
All Local Tribunals	159	2,658	1,793	2,281
Appellate Tribunal: (Decisions given on appeal)				
South England No.1 Division	12	348	100	168
South England No.2 Division	6	205	189	296
South England No.3 Division	2	334	152	99
North England Division	13	164	115	135
Wales Division	-	108	44	31
Scotland Division	4	169	45	97
All divisions of Appellate Tribunal	37	1,328	645	826

* Key to Decisions:

A Registration in COs register unconditionally.
B Registration in COs register conditionally upon undertaking civilian work or training.
C Registration in COs register as person liable to be called up but to be employed only in non-combatant duties.
D Name to be removed from COs register.

TABLE 3(e) 1943 Orders made by Tribunals on applications for registration in register of Conscientious Objectors under Section 5 of the National Service (Armed Forces) Act 1939 (Men)*

Tribunal	A	B	C	D
Local Tribunals:				
London No.1	12	197	236	439
London No.2	-	-	-	-
South Eastern	5	149	77	90
Southern	-	90	78	97
East Anglian	1	103	95	42
South Western	1	161	98	56
Midlands	5	169	35	64
North Midlands	-	43	23	42
North Eastern	2	73	29	53
North Western	6	96	77	77
Cumberland and Westmoreland	-	-	-	-
Northumberland and Durham	5	17	19	25
SE Scotland	3	13	31	48
SW Scotland No.1	20	11	16	79
SW Scotland No.2	-	22	18	95
N Scotland	1	-	1	3
NE Scotland	1	3	10	8
S Wales	8	115	56	43
N Wales	-	31	13	41
All Local Tribunals	70	1,293	842	1,302
Appellate Tribunal: (Decisions given on appeal)				
South England No.1 Division	5	178	51	78
South England No.2 Division	8	114	85	156
South England No.3 Division	-	148	43	62
North England Division	4	74	59	89
Wales Division	-	91	37	24
Scotland Division	7	81	46	94
All divisions of Appellate Tribunal	24	686	321	503

* Key to Decisions:

- A Registration in COs register unconditionally.
- B Registration in COs register conditionally upon undertaking civilian work or training.
- C Registration in COs register as person liable to be called up but to be employed only in non-combatant duties.
- D Name to be removed from COs register.

TABLE 3(f) 1944 Orders made by Tribunals on applications for registration in register of Conscientious Objectors under Section 5 of the National Service (Armed Forces) Act 1939 (Men)*

Tribunal	A	B	C	D
Local Tribunals:				
London No.1	12	79	86	165
London No.2	-	-	-	-
South Eastern	4	29	9	18
Southern	-	15	23	24
East Anglian	1	37	9	17
South Western	2	44	19	20
Midlands	9	127	17	40
North Midlands	-	43	16	38
North Eastern	2	37	10	30
North Western	5	52	31	45
Cumberland and Westmoreland	-	-	-	-
Northumberland and Durham	2	11	10	10
SE Scotland	4	21	11	20
SW Scotland No.1	24	22	19	84
SW Scotland No.2	-	1	1	7
NE Scotland	1	2	4	3
S Wales	4	30	16	11
N Wales	-	6	2	10
All Local Tribunals	70	556	283	542
Appellate Tribunal: (Decisions given on appeal)				
South England No.1 Division	1	47	36	41
South England No.2 Division	1	35	21	60
South England No.3 Division	-	58	24	23
North England Division	1	43	20	32
Wales Division	-	27	10	7
Scotland Division	-	25	15	53
All Divisions of Appellate Tribunal	3	235	126	216

* Key to Decisions:

A Registration in COs register unconditionally.
B Registration in COs register conditionally upon undertaking civilian work or training.
C Registration in COs register as person liable to be called up but to be employed only in non-combatant duties.
D Name to be removed from COs register.

TABLE 3(g) Cumulative total (1939-44)** Orders made by Tribunals on applications for registration in register of Conscientious Objectors under Section 5 of the National Service (Armed Forces) Act 1939 (Men)*

Tribunal	A	B	C	D	Total
Local Tribunals:					
London No.1	121	2,628	4,624	4,711	12,084
London No.2	75	667	527	1,025	2,294
South Eastern	214	2,024	1,432	1,862	5,532
Southern	45	1,112	1,003	971	3,131
East Anglian	268	1,329	682	320	2,599
South Western	563	2,598	1,169	560	4,890
Midlands	85	3,923	968	1,097	6,073
North Midlands	4	867	562	601	2,034
North Eastern	79	957	1,081	891	3,008
North Western	365	1,605	1,618	1,238	4,826
Cumberland and Westmoreland	30	99	218	115	462
Northumberland and Durham	108	403	281	286	1,078
SE Scotland	94	380	531	411	1,416
SW Scotland No.1	406	514	340	1,509	2,769
SW Scotland No.2	11	350	195	710	1,266
N Scotland	5	21	48	63	137
NE Scotland	16	196	150	97	459
S Wales	234	1,263	892	487	2,876
N Wales	76	952	394	423	1,846
All Local Tribunals	2,799	21,889	16,715	17,377	58,780
Appellate Tribunal: (Decisions given on appeal)					
South England No.1 Division	77	2,749	2,082	1,770	6,678
South England No.2 Division	36	1,555	1,233	1,411	4,235
South England No.3 Division	8	875	529	412	1,824
North England Division	106	859	621	729	2,315
Wales Division	3	407	172	120	702
Scotland Division	43	842	548	1,050	2,483
All Divisions of Appellate Tribunal	273	7,287	5,185	5,492	18,237

* Key To Decisions:

A Registration in COs register unconditionally.
B Registration in COs register conditionally upon undertaking civilian work or training.
C Registration in COs register as person liable to be called up but to be employed only in non-combatant duties.
D Name to be removed from COs register.

** Figures for 1945 unavailable.

TABLE 3(h) Other cases heard by Local Tribunals and the Appellate Tribunal*

	A	B	C	D	Total
Tribunals also gave decisions on 4,109 cases of conditionally registered COs referred by the Minister under Section 5 (1) of the National Service Act 1941. The Appellate Tribunal also heard 374 appeals against such decisions, 719 applications under Section 13 of the National Service (Armed Forces) Act 1939, by men in HM Forces sentenced by courts martial and 1,755 applications under Section 5 of the National Service (No 2) Act 1941 by men imprisoned for refusing to submit to medical examination. The resultant orders of the Tribunal are shown opposite	3,430	28,591	14,628	12,131	58,780
Women who were made liable for call up to the Women's Auxilliary Services by the National Service (No 2) Act 1941 were accorded the same rights as men to apply for registration in the register of COs and in their cases, up to the end of 1944, Tribunals made the orders shown opposite.	93	771	30	162	1,056
Total (men and women)	3,523	29,362	14,659	12,203	59,836

* Key to Decisions:

- A Registration in COs register unconditionally.
- B Registration in COs register conditionally upon undertaking civilian work or training.
- C Registration in COs register as person liable to be called up but to be employed only in non-combatant duties.
- D Name to be removed from COs register.

TABLE 4 Analysis of decisions of Local Tribunals under Section 5(1) to (3) of the National Service Act 1941 : cumulative figures up to and including 31 December 1948

Tribunal	No failure to comply with condition of registration	Failure to comply with condition of registration and no reasonable excuse	Failure to comply with condition of registration but with reasonable excuse and			Total
			No fresh order made	Appellant registered unconditionally	Condition of registration varied	
London No.1	1	19	23	84	1,935	2,062
S Eastern	-	-	-	9	136	156
Southern	2	15	1	2	261	281
E Anglian	-	5	4	8	217	234
S Western	10	3	4	72	424	513
Midlands	14	8	15	138	434	609
N Midlands	3	8	8	-	355	374
N Eastern	-	15	11	33	258	317
N Western	2	5	8	5	459	479
N Western (Cumberland and Westmoreland cases)	-	-	-	-	3	3
Northumberland and Durham	1	4	9	15	77	106
SE Scotland	6	10	4	23	102	145
SW Scotland No.1	23	56	9	247	283	618
SW Scotland No.2	9	6	3	20	94	132
N Scotland	1	-	-	1	5	7
NE Scotland	5	1	-	16	42	64
S Wales	13	14	-	41	270	338
N Wales	-	19	5	24	88	136
Men	89	187	104	716	5,351	6,447
Women	1	2	-	22	92	117
Cumulative total	90	189	104	738	5,443	6,564

TABLE 5 Conscientious Objectors registered conditionally upon undertaking civilian work - position regarding compliance with condition of registration

	Position on				
	31.7.40	31.3.41	31.3.42	31.3.43	31.3.45
Cumulative total no. of persons registered conditionally (excluding) those who since registration renounced their conscientious objection or enlisted voluntarily in HM Forces)	7,452	13,031	20,504	23,538 (501)**	24,625 (732)
Number known to be performing work of the kind ordered by the Tribunal or serving in Civil Defence	6,255	12,334	18,461	21,953 (312)	23,040 (583)
Number not performing work of the kind ordered:					
a No. regarded as having reasonable excuse, eg awaiting appeal, under submission for work, under reference back to Tribunal etc.	1,197***	697***	1,491	1,038 (127)	1,013 (65)
b No. under consideration with a view to prosecution			153	80 (17)	61 (15)
c No. in prison			19	11 (3)	7 (1)
Number where inquiries with a view to securing compliance were still proceeding			380	456 (42)	498 (68)
Cumulative total no. of persons prosecuted in connection with failure to comply with condition of registration	-	-	36	157 (11)	307 (84)

* Figures not available for 1944.

** Figures in brackets show, additionally, the no. of women.

*** Separate figures under b and c were not available for these dates which were prior to the operation of Section 5 of the National Service Act 1941.

TABLE 6(a) Appeals against decisions of Local Tribunals
Appellate Tribunal-Analysis of decisions given up to 31 December 1948

Division of Appellate Tribunal	No. of appeals heard	Local Tribunal decision A				Local Tribunal decision B				
1	2	3				4				
		Appellate Tribunal decision				Appellate Tribunal decision				
		A	B	C	D	A	B	B*	C	D
S England 1	6,875	-	1	-	2	21	259	281	109	6
S England 2	4,666	-	-	-	-	18	165	178	22	11
S England 3	2,034	-	-	-	-	3	132	86	7	2
N England	2,570	-	-	-	-	63	112	212	5	10
Wales	777	-		-	-	1	35	51	-	-
Scotland	2,732	-	-	-	-	14	123	123	1	-
All Divisions M**	19,224	-	2	-	2	114	735	830	144	28
All Divisions W	430	-	-	-	-	6	91	101	-	1
Cumulative total	19,654	-	2	-	2	120	826	931	144	29

* Nature of employment varied by Appellate Tribunal (included in column 7).

** M Men; W Women.

Division of Appellate Tribunal	Local Tribunal decision C				Local Tribunal decision D				Local Tribunal decision varied by AT	
	5				6					
	Appellate Tribunal decision				Appellate Tribunal decision				No. varied	%
	A	B	C	D	A	B	C	D		
S England 1	27	1,380	1,245	40	35	945	758	1,766	3,605	52.4
S England 2	9	814	705	53	15	605	549	1,522	2,274	48.7
S England 3	3	363	262	4	3	430	284	455	1,185	58.2
N England	29	463	435	14	18	204	222	783	1,240	48.2
Wales	1	214	127	-	2	146	56	143	472	60.7
Scotland	16	228	287	3	17	453	284	1,183	1,139	41.7
All Divisions M	85	3,449	3,055	114	84	2,678	2,148	5,756	9,678	50.3
All Divisions W	-	13	6	-	6	105	5	96	237	55.1
Cumulative total	85	3,462	3,061	114	90	2,783	2,153	5,852	9,915	50.4

TABLE 6(b) Appeals under Section 5(3) of National Service Act 1941 on question of compliance with condition of registration as Conscientious Objector. Appellate Tribunal - Analysis of decisions given up to 31 December 1948

Division of Appellate Tribunal		No. of appeals heard	Decision of Appellate Tribunal: No failure to comply with condition of registration	Decision of Appellate Tribunal: Failure to comply with condition of registration and no reasonable excuse
S England 1		76	-	5
S England 2		100	1	10
S England 3		79	1	8
N England		72	2	7
Wales		45	17	9
Scotland		113	21	10
All Divisions	M	478	42	49
	W	7	-	-
Cumulative total		485	42	49

Division of Appellate Tribunal		Decision of Appellate Tribunal: Failure to comply with condition of registration but with reasonable excuse and			LT decision varied by AT	
		No fresh order made (i.e. appeal dismissed)	Appellant registered unconditionally	Condition of registration varied	No. varied	%
S England 1		12	9	50	60	78.0
S England 2		25	5	59	67	67.0
S England 3		22	2	46	52	65.8
N England		21	2	40	47	65.3
Wales		6	-	13	30	66.6
Scotland		6	8	68	97	86.6
All Divisions	M	91	24	272	347	72.5
	W	1	2	4	6	85.7
Cumulative total		92	26	276	353	72.7

TABLE 7 Total number of appeals withdrawn up to 31 December 1948

Men	949
Women	9
Cumulative total	958

TABLE 8 Applications under Section 13 of National Service (Armed Forces) Act 1939 by men in HM Forces up to 31 December 1948

Division of Appellate Tribunal	No. of applications heard	Decision of Appellate Tribunal			
		Discharge from HM Forces not recommended	Discharge from HM Forces recommended and on discharge to be registered under		
			A	B	C
S England 1	133	23	-	105	5
S England 2	112	29	1	77	5
S England 3	96	4	-	86	6
N England	407	146	2	242	17
Wales	16	6	-	9	1
Scotland	44	13	2	24	5
All Divisions	808	221	5	543	39

TABLE 9 Applications under Section 5 of National Service (No.2) Act 1941 by men imprisoned for failing to submit to medical examination up to 31 December 1948

Division of Appellate Tribunal	No. of applications heard	Decision of Appellate Tribunal					Previous order varied by Appellate Tribunal	
		No conscientious objection found and no fresh order made		Conscientious objection found and applicant ordered			No. varied	%
		C	D	A	B	C		
S England 1	455	42	54	5	345	9	359	78.9
S England 2	499	98	118	2	214	17	233	51.8
S England 3	356	41	60	1	246	8	255	71.6
N England	395	79	137	4	157	18	179	45.3
Wales	27	5	4	-	18	-	18	66.6
Scotland	209	11	109	-	87	2	89	42.6
All Divisions	1,891	276	482	12	1,067	54	1,133	59.9

BIBLIOGRAPHY

1 MANUSCRIPT SOURCES AND PUBLIC RECORDS

Central Board of Conscientious Objectors Papers and Peace Pledge Union Papers, Friends House, Euston Road, London, N.W.1

Ministry of Labour and Home Office Papers, Public Record Office, Chancery Lane, London, W.C.2.
Ministry of Labour and National Service: Class 6, Pieces 2, 7, 12, 13, 14, 21, 107, 111, 125, 126, 127, 128, 137, 141, 142, 147, 148, 151, 165, 172, 183, 189, 207, 229, 264, 278, 282, 337, 405; Class 8, Pieces 179, 181, 193, 346, 475, 571, 604, 1041.
Home Office: Class 186, Pieces 616, 821, 822, 824, 913, 1147, 1155.
War Office: Class 32, Pieces 9432, 14529, Class 166, Piece 5831.
Cabinet: Class 23, Pieces 14(38)4, 15(39)5, 22(39)3, 25(39)4, 26(39)6.

House of Commons, Debates, Fifth Series, 1939-45.

Trades Union Congress Library, Congress House, London, W.C.1.

University Library, Cambridge.

2 INTERVIEWS

A number of informal conversations with conscientious objectors of the Second World War have been held. These provided most helpful and enlightening background material.

3 PERIODICALS

Press cuttings in the CBCO Papers at Friends House, Euston Road, London W1 and microfilm records at the University Library, Cambridge.

'Aberdeen Evening Express'
'Barnet Press'

'Bath Chronicle and Herald'
'Belfast News Letter'
'Berwick Advertiser'
'Birmingham Evening Dispatch'
'Birmingham Gazette'
'Bolton Evening News'
'Bournemouth Daily Echo'
'Brighton Gazette'
'Bristol Evening World'
'Carlisle Journal'
'Cooperative News'
'Coulsdon and Purley Times'
'County Advertiser and Herald'
'Country Herald'
'Country Life'
'Daily Dispatch'
'Daily Express'
'Daily Herald'
'Daily Mail'
'Daily Mirror'
'Daily Sketch'
'East Kent Gazette'
'Education'
'Empire News'
'Enfield Gazette and Observer'
'Evening Chronicle'
'Evening Standard'
'Farmers Weekly'
'Friend'
'Freethinker'
'Glasgow Herald'
'Hampshire Herald'
'Harrowgate Herald'
'Hemel Hempstead Gazette'
'Hertfordshire Mercury'
'Hornsey Journal'
'Kent Messenger'
'Law Journal'
'Manchester City News'
'Manchester Daily Dispatch'
'Manchester Evening Chronicle'
'Manchester Evening News'
'Manchester Guardian'
'Midlands Counties Tribune'
'Newcastle Evening Chronicle'
'News Chronicle'
'New Leader'
'New Statesman'
'Northamptonshire Evening Telegraph'
'Northampton Chronicle and Echo'
'Northampton Independent'
'Northern Daily Telegraph'
'Peace News'
'Preston Guardian'

'Reynolds News'
'Romford Recorder'
'Schoolmaster'
'Scotsman'
'Solicitor'
'Spectator'
'Star'
'Sunday Chronicle'
'Sunday Comet'
'Sunday Graphic'
'Sunday News'
'Sunday Pictorial'
'Sunday Referee'
'Teachers World'
'The Times'
'Weekly Review'
'Weekly Scotsman'
'Western Mail'
'Western Morning News'
'West Lancashire Evening Gazette'
'Widnes Weekly News'
'Wiltshire News'
'Worcester News and Times'
'World Review'
'Yorkshire Observer'
'Yorkshire Post'

4 AUTOBIOGRAPHIES, BIOGRAPHIES AND MEMOIRS

BEDFORD, SYBILLE (1973-4), 'Aldous Huxley: A Biography', 2 vols, London.
BEVIN, ERNEST (1942), 'The Job To Be Done', London.
BRITTAIN, VERA (1957), 'Testament of Experience', London.
BROCKWAY, FENNER (1947), 'Bermondsey Story: The Life of Alfred Salter', London.
BROCKWAY, FENNER (1942), 'Inside the Left: Thirty Years of Platform, Press, Prison and Parliament', London.
BROCKWAY, FENNER (1963), 'Outside the Right', London.
BROCKWAY, FENNER (1977), 'Towards Tomorrow: An Autobiography', London.
BULLOCK, ALAN (1960), 'The Life and Times of Ernest Bevin', vol.1, London.
CLARK, RONALD W. (1975), 'The Life of Bertrand Russell', London.
DAVIES, GEORGE M.Ll. (1950), 'Pilgrimage of Peace' (with an introduction by Charles Raven), London.
'Dick Sheppard: By his Friends' (1938), London.
DILLISTONE, F.W. (1975), 'Charles Raven: Naturalist, Historian, Theologian', London.
DONOUGHUE, B. and JONES, G.W. (1973), 'Herbert Morrison: Portrait of a Politician', London.
GILBERT, MARTIN (1965), 'Plough My Own Furrow: The Story of Lord Allen of Hurtwood as told through his own writings and correspondence', London.

GILL, ERIC (1940), 'Autobiography', London.
HASSLER, ALFRED (1955), 'Diary of a Self-made Convict', London.
INGE, WILLIAM RALPH (1939), 'A Pacifist in Trouble', London.
JAMESON, MARGARET STORM (1969-70), 'Journey from the North: Autobiography", 2 vols, London.
LEA, F.A. (1959), 'The Life of John Middleton Murry', London.
MARWICK, ARTHUR (1964), 'Clifford Allen: The Open Conspirator', Edinburgh and London.
MATTHEWS, C.H.S. (1948), 'Dick Sheppard: Man of Peace', London.
McNAIR, JOHN (1955), 'James Maxton: The Beloved Rebel', London.
MILNE, A.A. (1939), 'Autobiography', London.
MORRISON, HERBERT (1960), 'An Autobiography', London.
PARTRIDGE, FRANCES (1977), 'A Pacifist's War', London.
PARTRIDGE, FRANCES (1981), 'Memories', London.
PAXTON, WILLIAM, et al. (1938), 'Dick Sheppard: An Apostle of Brotherhood', London.
PETHICK-LAWRENCE Lord (1943), 'Fate Has Been Kind', London.
PLOWMAN, D.L. (ed.) (1944), 'Bridge Into The Future: Letters of Max Plowman', London.
POSTGATE, RAYMOND (1951), 'The Life of George Lansbury', London.
PURCELL, WILLIAM (1972), 'Portrait of Soper: A Biography of the Reverend Lord Soper of Kingsway', London.
ROBERTS, R. ELLIS (1942), "H.R.L. Sheppard: Life and Letters', London.
RUSSELL, BERTRAND (1951), 'Autobiography 1914-44', London.
SIMMONS, CLIFFORD (1965), 'The Objectors', Douglas, Isle of Man.
SPEAIGHT, ROBERT (1966), 'Life of Eric Gill', London.
TAIT, KATHERINE (1976), 'My Father Bertrand Russell', London.
THOMPSON, DOUGLAS (1971), 'Donald Soper: A Biography', London.
URWIN, E.C. (1955), 'Henry Carter C.B.E.: A Memoir', London.
WILLIAMSON, HUGH ROSS (1956), 'The Walled Garden: An Autobiography', London.
YORKE, MICHAEL (1981), 'Eric Gill: Man of Flesh and Spirit', London.

5 BOOKS AND ARTICLES

ALLEN, CLIFFORD (1930), 'Peace In Our Time', London.
ALLEN, E.L., F.E. POLLARD and G.A. SUTHERLAND (1946), 'The Case for Pacifism and Conscientious Objection: a reply to G.C. Field', London.
BAINTON, ROLAND H. (1916), 'Christian Attitudes Towards War and Peace: An Historical Survey and Critical Re-Evaluation', London.
BARKER, RACHEL (1978), Conscientious Objection in Great Britain, 1939-45, PhD Thesis, Cambridge.
BEDAU, H.A. (1969),'Civil Disobedience: Theory and Practice', New York.
BELL, JULIAN (ed.) (1935), 'We Did Not Fight: 1914-18 Experiences of War Registers', London.
BLISHEN, EDWARD (1972), 'A Cack-Handed War', London.
BOULTON, DAVID (1967), 'Objection Overruled', London.
BRITTAIN, VERA (1942), 'Humiliation With Honour', London.
BRITTAIN, VERA (1940), 'Letters to Peace Lovers', London.
BRITTAIN, VERA (1964), 'The Rebel Passion: A Short History of Some Pioneer Peace Makers', London.

BROCK, PETER (1970), 'Twentieth-Century Pacifism', New York.
BUTLER, D.E. and W.T. FREEMAN (1975), 'British Political Facts 1900-1968', London.
CAIN, EDWARD R. Conscientious Objection in France Britain and the United States, 'Comparative Politics', 2(2), January 1970, pp.274-307.
CADOUX, CECIL J. (1940),'Christian Pacifism Re-Examined', London.
CALDER, ANGUS (1969), 'The People's War', London.
CEADEL, MARTIN (1980), 'Pacifism in Britain 1914-45', London.
CHURCHILL, WINSTON L.S. (1948-54), 'The Second World War', London.
DAVIES, A. TEGLA (1962), 'Friends' Ambulance Unit: The Story of the F.A.U. in the Second World War', London.
FEINBERG, JOEL (ed.) (1969), 'Moral Concepts', London.
FIELD, G.C. (1945), 'Pacifism and Conscientious Objection', Cambridge.
GRAHAM, JOHN W. (1922), 'Conscription and Conscience: A History 1916-19, London.
HAYES, DENIS (1949), 'Challenge of Conscience: The Story of the Conscientious Objectors of 1939-45', London.
HOBBES, T. (1651), 'Leviathan', London.
HOWARD, MICHAEL (1977), 'War and the Liberal Conscience', London.
HUXLEY, ALDOUS (1972), 'An Encyclopedia of Pacifism', New York.
HUXLEY, ALDOUS (1937), 'Ends and Means: An Enquiry into the Nature of Ideals and into the Methods Employed for Their Realisations', London.
HUXLEY, ALDOUS (1936), Pacifism and Philosophy, in 'The New Pacifism', London.
JOAD, C.E.M. (1940), 'Journey Through the War Mind', London.
JOAD, C.E.M. (1939), 'Why War?', London.
LEWIS, JOHN (1940), 'The Case Against Pacifism', London.
LUARD, EVAN (1962), 'Peace and Opinion', London.
MACAULAY, ROSE, et al. (1937), 'Let Us Honour Peace', London.
MACGREGOR, G.H.C. (1936), 'The New Testament Basis of Pacifism', London.
MARTIN, DAVID A. (1965), 'Pacifism: An Historical and Sociological Study', London.
MACNEIL, HECTOR, Foreign Policy Between the Wars in 'The British Labour Party', vol.2, ed., Herbert Tracy.
MARWICK, ARTHUR (1976), 'The Home Front: The British and the Second World War', London.
MAYER, PETER (ed.) (1966), 'The Pacifist Conscience', London.
MELLANBY, KEITH (1945), 'Human Guinea-Pigs', London.
MILNE, A.A. (1934), 'Peace With Honour: An Enquiry Into the War Convention', London.
MILNE, A.A. (1940), 'War With Honour', London.
MORGAN, KENNETH O., Peace Movements in Wales, 1899-1945, 'Welsh History Review', vol.10, no.3, 1981.
MORRIS, DAVID (1948), 'China Changed My Mind', London.
MORRISON, SYBIL (1962), 'I Renounce War: The Story of the Peace Pledge Union', London.
MUMFORD, PHILIP S. (1937), 'An Introduction to Pacifism', PPU.
'Municipal Yearbook', 1940 edition, London, 1897-.
MURRY, JOHN MIDDLETON (1957), 'Community Farm', London.
MURRY, JOHN MIDDLETON (1937), 'The Necessity of Pacifism', London.
MURRY, JOHN MIDDLETON (1938), 'The Pledge of Peace', London.

PARKER, H.M.D. (1957), 'Manpower: A Study of War-time Policy And Administration', History of the Second World War: U.K. Civil Service, London.
PELLING, HENRY (1970), 'Britain and the Second World War', London.
POLLARD, FRANCES E. (1953), 'War and Human Values', London.
POLLARD, ROBERT S., Conscientious Objectors: Great Britain and the Dominions, 'Journal of Comparative Legislation and International Law', Third Series, vol.28, 1946-7.
PONSONBY, LORD (ARTHUR) (1925), 'Now is the Time: An Appeal for Peace', London
PRASAD, D. and T. SMYTHE (1968), 'Conscription: A World Survey', London.
RAE, JOHN (1970), 'Conscience and Politics: The British Government and the Conscientious Objectors to Military Service 1916-19', London.
RAVEN, CHARLES E. (1938), 'War and the Christian', London.
RICHARDS, LEYTON (1937), 'The Christian's Alternative to War: An Examination of Christian Pacifism', London.
ROBBINS, KEITH (1972), 'The Abolition of War: 'The Peace Movement' in Britain 1914-19', Cardiff.
RUSSELL, BERTRAND (1946), 'History of Western Philosophy', London.
RUSSELL, BERTRAND (1916), 'Justice in War Time', Manchester.
RUSSELL, BERTRAND (1936), 'Which Way to Peace?', London.
SHAW, G. BERNARD (1944), 'Everybody's Political What's What', London.
SHAW, G. BERNARD (1948), The Unavoidable Subject, 'Journal of the War Years 1939-45 and one year later', London.
SHAW, G. BERNARD (1931), 'What I Really Wrote About The War', Edinburgh.
SINGER, P. (1973), 'Democracy and Disobedience', Oxford.
STRATMANN, FRANZISKUS (1928), 'The Church and War: A Catholic Study', London.
TAYLOR, A.J.P. (1965), 'England 1914-45', Oxford.
TOLSTOY, LEO (1967), 'Writings on Civil Disobedience and Non-Violence', New York.
VAN DEN HAAG, ERNEST (1972), 'Political Violence and Civil Disobedience', New York.
WATT, W.M. (1937), 'Can Christians be Pacifists?', London.
WEATHERHEAD, LESLIE (1939), 'Thinking Aloud in Wartime', London.
WHITE, L.E. (1946), 'Tenement Town', London.
'Who's Who' (1970), London.
'Who Was Who', 1941-50, 1951-60, 1961-70, London.
WITTNER, LAWRENCE S. (1969), 'Rebels Against War: American Peace Movement 1941-1960', New York.
WILSON, ROGER C. (1952), 'Quaker Relief: An Account of the Relief Work of the Society of Friends 1940-48', London.
ZAHN, G.C. (1967), 'War, Conscience and Dissent', London.

6 PAMPHLETS AND BROADSHEETS PUBLISHED BY THE CBCO
(All published in London)

Conscription: A Commentary on the National Service (Armed Forces) Act, 1939.

A Description of the Military Training Act, 1939.
Court Martial Guide and Friend, 1940.
For Your Urgent Attention, 1940.
Enlistment for Military Service, 1941.
Police Court Procedure, 1941.
The C.O. and the National Service Acts, 1942.
Cat and Mouse, 1942.
Compulsory Fire Watching at Your Place of Work, 1942.
Conscription for war work, 1942.
Enrolment for Civil Defence, 1942.
Women and Military Service, 1942.
Duties on Civil Defence, 1943.
Fire Guard Duty Under Your Local Authority, 1943.
Medical Examination, 1943.
Work on the Land, 1943.
Call-up for the Home Guard, 1944.
Part Time Civil Defence, 1944.
The Present Position of Conscientious Objection, 1944.
The C.O. and the Future, 1945.
Control of Employment, 1945.
Freedom of Conscience by Robert S.W. Pollard, 1945.
The Flowery (The Scrubs 'Conchie' Review), 1945.
Non-fulfilment of a Condition, 1945.
C.O.'s in Great Britain, 1950.
C.O.'s - Their Position in 1953, 1953.
The Unconditionalists, 1954.
Non-combatant Duties, 1955.
The Chaplains Guide to C.O.'s in the Armed Forces, 1956.
Civil Prisons: A Guide to Conditions, 1957.
A Short History of the Legal Position of Conscientious Objectors to Military Service and Other War-Time Compulsions by Constance Braithwaite, 1964.
Notes on Prison Routine, undated.
National Service: A Guide for the C.O., post war.
National Service and Registration as a C.O., post war.
Catalogue of Conviction: George Elphick, undated.
The C.O. and the Community by Fenner Brockway, undated.
The C.O. and the Tribunal, undated.
Registering as a C.O., post war.
The Release of C.O.'s, post war.
To C.O.'s Wishing To Appeal, undated.
On Being a Good Witness, undated.
Proceedings of the Appellate Tribunal, undated.
Civil and Military Prison Routine, undated.
Questions to C.O.'s, undated.
The Case for Stanley Hilton, undated.

7 THESES AND DISSERTATIONS

HUGHES, JOHN, The Legal Implications of Conscientious Objection, Master of Laws, Manchester, 1971.
WINTER, J.M., The Development of British Socialist Thought 1912-18, Doctor of Philosophy, Cambridge, 1970.

INDEX